"So You're the Safety Director!"

An Introduction to Loss Control
and Safety Management

Second Edition

Michael V. Manning

Government Institutes, Inc.
Rockville, Maryland

Government Institutes, Inc., 4 Research Place, Rockville, Maryland
20850, USA.
Phone: (301) 921-2300
Fax: (301) 921-0373
Email: giinfo@govinst.com
Internet address: http://www.govinst.com

02 01 00 99   5 4 3 2

The reader should not rely on this publication to address specific
questions that apply to a particular set of facts. The author and publisher
make no representation or warranty, express or implied, as to the
completeness, correctness or utility of the information in this
publication. In addition, the author and publisher assume no liability of
any kind whatsoever resulting from the use of or reliance upon the
contents of this book.

Library of Congress Cataloging-in-Publication Data

Manning, Michael V., 1943.
So you're the safety director! : loss control and safety management, 2nd
ed. / Michael V. Manning.
        p.   cm.
    Includes bibliographical references and index.
    ISBN: 0-86587-617-7
    1. Industrial safety--Handbooks, manuals, etc.   I. Title.
T55.M3516   1998
658.4'08--dc20
                                95-43644
                                CIP

Printed in the United States of America

Dedicated To:

You, the reader—what you do is significant and valuable!

John Gross, a CEO who understands that the safety of his employees is equally as important as the production of his product.

Lynn, who keeps me off the wrong bus.

—Michael V. Manning, Ph.D.

ABOUT THE AUTHOR

Dr. Michael V. Manning, President of Manning & Associates, specializes in Safety Engineering and Education. Dr. Manning has more than 30 years in the safety profession serving as a Senior Engineer for a national insurance company, a Safety Engineer for NASA, and as a Corporate Risk Management Director for an international corporation. He is the author of the books, "So You're the Safety Director" and "Safety Is a People Business." Michael writes for several national safety publications, conducts seminars nationwide on the formulation and implementation of safety programs and provides safety management services for companies.

Dr. Manning holds a Ph.D. in Occupational Safety and Health, is a Certified Safety Executive with the World Safety Organization, and is an Associate Safety Professional with the Board of Certified Safety Professionals. He is a member of the American Society of Safety Engineers, the President of the Associated Safety Engineers of Arizona and is an Adjunct Professor for Columbia Southern University.

TABLE OF CONTENTS

Appendices / 153

FOREWORD

So You're Not a Safety Professional?

This manual is not intended for safety professionals. It is for those who have had the responsibility of a safety program thrust upon them. There is no complete substitute for the qualified safety professional in developing, implementing, and measuring the effectiveness of a comprehensive safety program. It is possible, though, for a draftee to build a viable framework. This manual is intended for that purpose.

In the thirty years I have spent in the safety profession, seven and a half were as a loss control consultant for a large insurance company. In my work I called on all types of businesses to assess the loss control methods they had in place and to offer my assistance in loss reduction. In areas from workers' compensation to product liability, I found one common denominator—the vast majority of those accounts did not have a full-time Safety Director, but rather the task was an add-on assignment for someone in accounting, personnel or other functional area.

Safety professionals have thousands of sophisticated references, manuals and guides to select from. But the nonprofessionals I observed were not taking advantage of the tools available to develop and implement the common components of a safety program. I would sit in their offices and ask to see their OSHA logs, loss runs, policies and programs. Their reactions would vary from embarrassment to defiance.

This manual is written for the new safety director who needs a primer or a cookbook. The very first thing you need to know is that you don't have to reinvent the wheel. Almost all of what you need has already been done for you. Remember that you are not alone. There are many resources as close as your telephone, and you are already paying for some of them.

A particularly useful resource for new Safety Directors is the Appendix section of this book, which contains sample forms and guidelines for almost every purpose.

In the final chapter of the book, I include the formal safety program that I developed and used with a mid-sized corporation in its two plants. I reduced their workers' compensation losses by 84 per cent and did a solid job of complying with state and federal standards within a five-year period. I did not do it alone, but rather I put together a team that was largely responsible for my success.

It is important for you to recognize and use the team approach. Take a moment to look through the Safety Program in Chpater XIV. It may look much more impressive than it really is. Think of it as a blueprint. In many instances, the key to your success will be recognizing whether you have to do something yourself or need to pick up the phone and have someone else do it.

Some Terminology

The best definition of "safety" I can find comes right out of the dictionary: freedom from danger, risk, or injury. "Risk" can be defined as the possibility of harm or loss.

Many safety-related job titles have been developed over the past twenty years. The "Safety Director" is now the loss prevention director, loss control administrator, safety director, company safety engineer, etc. To make this manual as down-to-earth as possible, I refer to your program as the "safety program" and to your job as the "Safety Director."

You Are Important!

As the Safety Director, you are responsible for protecting your employees and the public from risks that could cause them harm. You are charged with protecting the assets of your company and with achieving compliance with federal, state, and local regulations. Most importantly, you are also charged with a moral responsibility to protect employees from injuries and fatalities. Every day, twenty-nine American workers are killed on the job—every day! Every day, four thousand seven hundred are injured so seriously that they must be taken to a medical care facility—every day! (National Safety Council, Accident Facts). What you do as the Safety Director is valuable. Others who do not understand your job may not agree simply because they don't know what it is you do. Tell them! Show them! Demonstrate by words and actions so they will be able to assist you in your efforts. Your responsibilities as the company Safety Director are meaningful, significant and valuable.

Real World: An Introduction

Wouldn't it be nice if it were a perfect world and everyone followed a "talk straight, play fair" organizational philosophy? You and I know that isn't true. At the end of each chapter is a section entitled "Real World" to address organizational players who won't, who don't, and who can't.

It is easy for me to write about my experiences as a Safety Director and how I was able to accomplish what I did. Your situation is probably different from mine. In your real world, you will need to address "blockers" (those people who for whatever reason stand in your way) and other problems that are different from the ones I have dealt with.

Where Does Your Authority Come From?

Usually what I found when visiting businesses is that the Safety Director's job would be assigned to someone (and in some cases, anyone) who would or could take it. The company could then tell their insurance agent, insurance carrier, OSHA, and all other federal and state compliance agencies that they have a Safety Director and a safety program.

The scope of the job is very wide. The Safety Director is involved in every area of the company, yet this person bumps into a lot of other people with higher status. In the real world, just because you tell a department supervisor that his or her employees need to wear eye protection doesn't mean the department will comply. You may be told it's not needed, or the supervisor will go to the vice president of production to complain that the "new Safety Director is slowing down the production line." Or you may simply be ignored. If you don't have the authority and backing to do the job, you will be a mild irritant at best and an amusing interruption at worst.

Talk to your boss or the person who assigned you the Safety Director role. Find out just how much authority you really have and how far you can push before you must stop. Clarify what the consequences are for a supervisor, employee, or even a department head who will not do what you say they must do. And be sure you understand who will be backing you up when you hit choppy water. It is not unusual that you will be told, "The company cares about its employees, and overall we want a safe work place. However, we will need to look at each situation individually." Look out—you are being told that the safety program is important to a point, but don't get carried away. This is doubletalk. OSHA, EPA, your insurance company and all the state and federal agencies can "get carried away" but not you? Nonsense!

A Safety Director must have the authority to run a safety program. Ask, How sincere—really sincere—is the company in starting this program? What do you really want me to accomplish? How will I demonstrate that I have met your expectations? Show your boss this manual and discuss the duties and responsibilities of the Safety Director and the procedures outlined in the following chapters. Is this what your boss wants you to do? (In some cases an "on-paper" program is really all that is wanted; if so, find out what boundaries you will be restricted to.)

Learn Your Company

Do some hard thinking about how to work within the company's organization. Ask yourself, How much do I know about this company, and how much experience do I have with it? Do I understand about the day-to-day operations of my company?

Take a tour from receiving dock to shipping dock. Learn as much as you can about how each department affects the production of the final product. If possible, have each department's supervisor go with you and show you the operation so you can see how it fits into the big picture. If possible sit down with each department supervisor to tell him or her your new role and the expectations your boss has of you. Explain that you are there to help make the department free from unsafe conditions and unsafe acts. Ask what the supervisor's safety problems and frustrations are in reducing injuries.

Don't be disappointed if the supervisor doesn't jump for joy at your arrival. Many supervisors will associate your role of Safety Director as meaning extra responsibilities for them.

Learn to Be Assertive

Evaluate your own personal style. Supervisors and employees will give you a million reasons why safety rules can't be enforced. You will have to draw a line, dig in and not give up. A Safety Director in the real world must be assertive. You are going to need all the interpersonal skills you can muster, but you are also going to have to be firm and strong. For some Safety Directors that is not easy. Evaluate how assertive and tenacious you are now. Decide how you can develop the reputation for authority that you will need. Also plan to work on interpersonal skills to exercise in situations where you must improvise your responses constantly.

No matter how noble the task of the safety profession, people usually view Safety Directors with a wary eye. *Safety Directors "cause work, hold people accountable, and find something wrong."* Safety Directors are paid to say "what if." You will never be able to please all the people, no matter how hard you try. Frustration can be the constant companion of the Safety Director if you let it. Remind yourself that what you do does make a difference.

Chapter I

"What am I supposed to do?"

The Basic Tasks of the Safety Director

The first task of the Safety Director is to **identify** the risks to your company. This means that you will:

- Develop methods of identifying risks.
- Identify the economic loss to your company.
- Review and interpret safety reports and other pertinent information.

The second task is to **develop and implement** a safety program. This means that you will:

- Develop policies, procedures and specific programs to eliminate causative factors.
- Review your safety program to ensure that it is current and responsive to needs.

The third task is to **measure** the effectiveness of your program. This means that you will:

- Develop measurement tools to determine effectiveness of the program.
- Develop measurement tools to determine cost reduction.

The three tasks of Identifying the Risks to Your Company, Developing and Implementing a Safety program, and Measuring the Effectiveness of Your Program are the foundation, the basis of what it is you do in your role. However, let's get a little more specific on what all that means.

Specific Duties and Responsibilities of the Company Safety Director

1. *Coordinate safety activities throughout the company.* That means exactly what it says. You, Safety Director, coordinate *all* safety activities in your company. You will be the person, the technical expert, that others look to for advice and recommendations.

2. *Maintain and analyze all accident reports for thoroughness and clarity.* In subsequent chapters we cover the information you will need for this task. Your insurance loss run, OSHA log, and incident rate will give you that data. Always pretend that the company President will be asking you the who, what, when, where, how, and why of injuries occurring in your company. You should be prepared to answer those questions.

3. *Determine and coordinate safety education for company employees.* Safety training of employees is the most significant and critical part of your job! Not only are you responsible for determining what training is needed, but how it will be delivered. Your employee's safety training should not only be OSHA mandated training, but other training that will increase your employees knowledge to minimize unsafe acts and unsafe conditions.

4. *Coordinate activities to maintain interest of company employees using bulletin boards, posters, and other informational transmitting methods as needed.* As the Company Safety Director you are required to be your program's communicator. You must be selling safety in whatever method works for you and your program.

Imagine yourself in a sandwich board with a safety message on the front and back. Your employees should see you in that role. Again, the success of your program is in many ways based on your ability to communicate.

5. *Coordinate accident investigations and near miss reports.* The book's chapter on Accident Investigation will give you the foundation for the critical task. You are the company representative to ensure that your supervisor's are doing the type of work that will prevent similar accidents from recurring. You are responsible for reviewing their reports for accuracy and preventive techniques.

6. *Coordinate all internal safety inspections.* One important method of discovering unsafe conditions, and often eliminating unsafe acts, are company safety inspections. The book's chapter on this topic will again give you suggestions on how to conduct a safety inspection. Certainly, it is the primary responsibility of your department supervisors, but you too must get out of your office and be seen conducting your own inspections.

7. *Ensure current knowledge of OSHA standards.* You can't play the game if you don't know the rules. It is "homework" time for you in this role. You must learn how to research and interpret OSHA standards. Again, employees will see you as the company's technical expert.

8. *Issue reports showing accident/incident patterns and trends.* Where are the injuries increasing or decreasing? What departments have the highest incident rate, lost time rate, etc. Your patterns and trends can often be obtained from your "roadmap" to injuries—your OSHA 200 log.

9. *Serve as the companies contact with your insurance carrier.* Who better to liaison with your insurance loss control inspector and claims representative than you? It's your safety program they are working with and you should be the person to explain it to them.

10. *Serve as the primary contact with OSHA compliance officers.* Later on in the book we have a chapter on Surviving An OSHA Inspection. It is a chapter you will need to read several times. You

are the contact person, and it will be you the compliance officer will be asking to see. You are the best representative of your company and its safety program. Know what to do and what not to do!

11. *Maintain current recordkeeping and training records as mandated by OSHA*. Again, you are the primary contact with OSHA. It will be you they are asking for these records. It will be you that will be, or should be, explaining what training and records you are keeping for your program. Who else should be responsible for these records, but you?

12. *The company Safety Director is responsible to keep the company president briefed and informed on the status of the safety program.* You are the person to keep, not only the President, but all senior management as to the current status of your program. You can do this verbally or by a monthly or quarterly report. The absolute worst time to inform the President of the status of your program is after an OSHA inspection. Let others in senior management know what it is you are doing and what it is you need.

These duties and responsibilities sound overwhelming. But don't worry. By following the safety program outlined in this text, you will learn how to accomplish them.

Achieving Those Safety Goals

So many times when a safety position is being developed in a company, people will have absolutely no idea what the role demands. You may hear comments like "Oh, you're responsible for checking the fire extinguishers," or "You're responsible for keeping the plant clean, eh?" Don't let these stereotypes bother you. Prior to your assuming this role you probably didn't realize the importance of the Safety Director either. (In the beginning of my career as a Safety Director I literally made copies of my job description, and when people would ask what it was that I did, I simply handed the job description to them, and smiled).

There are going to be those who will minimize your role in the safety position no matter what approach you take—it is a fact of life, particularly in small businesses. But at least you will know the significance of your position, and your boss will know as well!

Funding the Safety Program

When you are challenged with creating a safety program and putting together a team, you and your boss are not going to know what the dollars and cents cost will be. That's okay. You will have to feel your way at first until you get a budget baseline. Here are some tips to get you started.

1) Find out the total cost of your insurance by coverage. It will be easier to request funding when you can compare your request to money already being spent.

2) Contact your local safety council, trade associations, and commercial course providers and publishers for available courses and books.

3) Don't forget the critical importance of training your supervisors. Again, contact your local safety council, trade associations, commercial course providers and publishers to get information and prices.

4) Realize that clerical staff for memos and agendas is going to be a "feel as you go" situation. Many of your memos can be handwritten, but the safety committee agendas need to be typed.

5) Sit down with your boss and explain what you plan to do to come up with a budget framework. As time goes by you will be able to be more precise as to the costs. For now, use the cost of your total insurance divided by the number of employees to push home the point that the safety program relative to insurance cost is a good investment.

You will note that I will frequently mention the National Safety Council (NSC), Telephone number 1-800-621-7619. There are many firms that can provide you with the basics, but I have found the NSC to be the best for those in your position who are just starting a program.

Chapter II

"Do we have a policy on that?"

The Administrative Process and the Safety Policy

Nothing can be so confusing and overwhelming to the Safety Director as the formulation and implementation of safety policy statements. Much of our consulting is involved in advising and assisting companies in that process. The one common denominator that I find is this: Safety Directors' attempts to make it much more difficult than it really is. Let's first look at the basis of the policy statement.

The Safety Policy Statement

Your safety policy statement is the keystone of your company safety program. In your policy statements you will be stating what is and what is not permitted to protect your employees from death and injury as well as being in compliance with state and federal safety mandates. What I often encounter with companies is that they either write their policies with such depth and length that it makes the OSHA standards look like a brief article. Other companies have their policies so short and so condensed that their employees, and more importantly, their first line supervisors spend the majority of their time scratching their heads as to what the intent is. There really is a happy medium—yes, there really is a workable happy medium. With all this in mind let me list some questions for you to ask yourself as your start writing yours.

1. What is it really that you intend to achieve with your policy statement? In many instances I find that the Safety Director is writing them because it's something they are supposed to do, it's just part of the safety program. Well, yes it is an intricate part of your program, but you must have a goal in mind. So again, determine what it is you want to achieve. Spend some time on this before you put pen to paper.

2. What are the next areas you wish to cover with your policy statements? My first suggestion in this area is to start with OSHA mandates and then work out from that. Areas covering Off-The-Job Safety, Vehicular Safety, Parking Lot Safety, etc., should come after your mandated standards.

3. The next big question is who is responsible for ensuring that these policy statements are enforced? This is one of two most important questions you should be asking yourself. Yes, you and senior management are, of course, ultimately responsible for seeing that the policies are carried out. That being said, it will almost always be your first line supervisors who will be enforcing these rules, these policies you are writing.

4. The next and last big question to ask yourself is how will individuals be held *accountable* for following the policy statements? Wherever I go I find companies with outstanding policy statements with absolutely no method, or intent for that matter, for enforcement. They are part of what I call "Hub Cap" safety programs. They look good but they simply provide no purpose for the protection of employees.

Writing the Policy Statement

Let's now get into writing the policy statements themselves. Remember, you should write one policy at a time. Don't get overwhelmed, and don't try to create the new American novel. As I stated before start with the mandate OSHA standards. Let's start with the OSHA standard on "Walking and Working Surfaces." It is certainly a crucial element of your requirements for a safe workplace. If you scan the standard you will see all the requirements you will be held accountable for. However, if you get the general intent of the standard you can then break it down into a workable policy statement. Let's now start writing the policy statement a line at a time.

YOUR COMPANY
Date: (The date the policy was issued.)
Subject: Safety—Walking and Working Surfaces
Purpose: To Comply with OSHA Standard 1910.22 (*Always put a purpose in your policy statement. What is it you are trying to achieve? Also, state the relevant OSHA standard by number in your policy statement. Many Compliance Officers have told me, after reading policies we have written for companies, that this indicates to them that the company is aware of their safety responsibilities and the relevant standard.*)

I. General Requirements
    A.   All work areas, including passages and storage areas will be maintained in a clean and orderly manner and free of all debris. (*You don't have to list all the requirements of the standard, get the salient points and include them.*)
    B.   The floor of all work areas will be maintained in a dry condition.
    C.   All permanent aisles will be maintained.
    D.   All permanent aisle ways will be kept free and unobstructed for vehicular and pedestrian traffic.

II. Administrative Requirements: (*You can label this anyway you want as long as your employees understand it is telling them who is responsible and accountable for it's compliance*).

A.   All first line supervisors and managers are responsible for the enforcement of this policy. (*You should put the relevant title of those people who are required to enforce this policy.*)

B.   All employees of YOUR COMPANY are required to obey the requirements of this safety policy. All employees are responsible and accountable for maintaining safe Walking and Working Surfaces. (*Make everyone in the company responsible and accountable for safety.*)

C.   Working safely is a condition of employment at YOUR COMPANY. (*I always put this at the bottom of the policy statement and recommend that you do to. This is a remainder to your employees that working safely is a company work rule, just like attendance or no drugs or alcohol.*)

Signature_____**,Title**_____

(*Have the senior official at your company sign the policy. Many companies with corporate headquarters issue out their edicts with some stranger's name on the bottom. It really means very little to them. Having the senior official allows the employees to put a face with the signature.*)

What do you think? Will it work for you? If so, let's now write another policy from the OSHA standards. Look up the standard 1910.132, the OSHA requirement on Personal Protective Equipment. Read through it and determine what it is you want to put in your policy statement. Please understand that with a standard this important you certainly need a policy statement. You don't need policy statements for each OSHA standards, but the most cited ones, and the ones that require an activist participation by your supervisors, you must. This is a long and involved standard and it will probably be at least two pages for your policy statement. Now take some time and pull out the important points of what it is you want in your policy statement.

YOUR COMPANY
Date:
Purpose: To Comply with OSHA Standard 1910.132

I. Application (*I used the term "application" right from the standard. If the policy was small I would have probably used "general requirements." It's okay for you to label your headings any way you want, it's your policy statement.*)

    A.   Protective equipment will be provided, used and maintained in a sanitary and reliable condition.

    B.   Protective Equipment will be used wherever it is necessary due to hazards associated with processes or environment, chemicals, mechanical irritants that may be present in the workplace.

II. Employee-Owned Equipment

    A.   Precautions will be taken by the company to ensure protective equipment provided by employees is adequate.

    B.   Precautions will be taken to ensure that employee-owned equipment is properly maintained and sanitized.

III. Design

    A.   All personal protective equipment will be of safe design and construction for the work to be performed.

IV. Hazard Assessment and Equipment Selection

    A.   The company will examine the workplace to determine if hazards are present, or are likely to be present, which would require the use of PPE.

    B.   The company will select, and have each affected employee use, the types of PPE that will protect the affected employee from the hazards identified in hazard assessment. The company will communicate selection decisions to each affected employee and select PPE that properly fits each affected employee.

    C.   The company will provide written proof that the required workplace hazard assessment has been performed, the person certifying that the evaluation has been performed and the date(s) of the hazard assessment and which identifies the document as a certification hazard assessment.

V. Defective and Damaged Equipment

    A.   The company will ensure that defective or damaged Personal Protective Equipment will not be used.

VI. Training

    A.   The company will provide training to each employee who is required to use PPE.

B.   Each employee will be trained in the following:
1.   When PPE is necessary.
2.   What PPE is necessary.
3.   How to properly put on, take off, adjust and wear PPE.
4.   The limitations of the PPE.
5.   The proper care, maintenance, useful life, and disposal of the PPE.

C.   Each affected employee will demonstrate understanding of the training.

D.   Each employee will be retrained where there is reason to believe that the previously trained employee does not have the understanding and skill required.

E.   Retraining will be conducted when:
1.   There are changes in PPE
2.   There are changes in exposure in the workplace

F.   The company will provide written proof that each affected employee has received and understood the required training.

VII.   Administrative Requirements

A.   All First Line Supervisors and Managers are responsible for enforcement of this policy.

B.   All employees of YOUR COMPANY are required to obey the conditions of this policy.

C.   All employees are responsible and accountable for following mandated company and OSHA regulations on Personal Protective Equipment.

D.   Working safely is a condition of employment at YOUR COMPANY.

Signature_____**Title**_____

Now, read over that policy statement. Does it look a lot like it came right out of the OSHA standard? It should, it was paraphrased directly from the CFR. Why go to all the time, work, and expense to develop elaborate safety policy statements when OSHA has laid out the requirements for you? You certainly don't need to include everything listed in the standard, but by listing the critical points you will have a policy that is workable.

The Administrative Process

Writing the safety policy statement really is the easiest part of the administrative process. I recommend a five-step process that we use successfully in companies all over the country. Let's see if it will work for you.

1. Write your policies and place them in a binder. Be sure to include your relevant OSHA mandated standards first. Then you can begin to add your internal policies. You don't have to write all your policy statements, but have enough to start you binder.

2. Take your policy statements to your safety committee for their review and your explanation. Be very clear on the role of the committees involvement. You are the Safety Director and you are responsible for ensuring the safety of your employees. That being said, allow the committee to have input and review of all of them for relevancy and applicability. One point you should make with the committee that is crucial to what you are trying to accomplish: "Let's be very clear about what we want, because this is the only way we will do things from now on." You wrote the policy statements to have a policy, a course or method of doing things. Again, be sure that those that need input get it, then move on.

3. After the Safety Committee show the policies to your first line supervisors for input. They are the ones that have to enforce what you have written. They certainly have the right to make their recommendations and suggestions. Once you get their input then it's time to start the use of the policies.

4. In this step the department supervisor takes his employees through the policies. This can be accomplished on a weekly basis or quarterly training. Whatever system they use, be sure that the supervisors educate the employees on the policies. Your safety policies are again, the keystone of your program.

5. The final step is the most important. The policies are worthless unless they are followed. You as the Safety Director need to ensure

that the policies you write are enforced by the department supervisors.

Your safety program will fail when administrative control fails. Administrative control always fail when your safety policies are not enforced. Your safety policies are your employees' way of acting or proceeding to ensure consistent performance. This is a critical step to ensure responsibility and accountability among your employees. Working safely truly is a condition of employment.

Employee Safety Rulebook

We have covered policies, let's now spend some time on your employee rulebook. In the appendix of this book is an Employee Safety Rulebook (Appendix F) that I wrote and used with companies nation wide. It is generic in nature and you should feel free to make as many changes as you deem necessary. The basics of what you want to include in your employee handbook is there for you to use. Let's go through it section by section.

First Page
There are three general sections:
1. The explanation of the philosophy of your company.
2. Three safety statements your company believes in and will support.
3. The signatures of the most important individuals in your program: The company president, the operations manager, (the person who is in charge of the people you have to protect) and you! It is these three people who generate the safety program. Have their names and signatures on this employee rulebook.

Second Page
This page contains your company instructions in the safety rulebook. It should be the very next page of your rulebook. This statement gives instructions on the use of the safety rulebook and the accountability of the employee. Notice that it is signed by the safety committee. These are the people who meet monthly to make your plant a safer place. Have them involved in this process.

Know Your Job

Unsafe Conditions—This is both an invitation and instructions to your employees to report what they see. Start immediately to get your line employees involved in your safety process.

Complete Instructions—Be very clear about the fact your employees must know what it is they are doing before they attempt to do it. Emphasize this point strongly!

Housekeeping—This is included because one of the major causes of death and injury in the workplace are slip, trip, fall, hazards caused by poor housekeeping. Set a standard for your employees and enforce it.

Safety Shoes—In the event your company requires the use of foot protection then include it in your rulebook. State how the employee will be reimbursed for their cost.

Eye Protection—In almost any type of industrial or construction exposure eye protection is a must. On an average year there are 140,000 eye injuries. Be specific and firm about this topic.

Gloves—Hand injuries in many cases can be prevented by hand protection. Insist that your employees wear them.

Fork Trucks—Forklifts are a great exposure to your employees, chiefly your new employees who have not had the experience of the industrial setting. Stress this point.

Safe Tools—Emphasize the need for safe and properly adjusted tools. Make it the employee's responsibility to check his/her tools daily.

Guards—Stress this point to the extreme. Ensure that your employees understand they cannot operate equipment without guards.

Electrical Repairs—The point you want to make is that only authorized service personnel are permitted to perform electrical work. Regardless of whether or not they are weekend electricians, they are not permitted to perform the work.

Air Hose—Many safety rulebooks ignore this point. OSHA standards prohibit air lines to exceed 30 PSI. Do not allow employees to use air lines to blow off clothing or hair.

Miscellaneous—This is the area where you can include what is specific to your company. Feel free to include whatever you consider relevant. I suggest you go back over your OSHA log for the past few years to include those accidents that were unusual and not in the exposure norm.

Fire Rules—Spend time on this section. Include sections from your emergency action plan. This is one of the most important parts of your safety rulebook. Do not waste this opportunity.

Conclusion—This states what the company is doing to protect your workers. It also states what their responsibility is. More importantly it requires them, your employees, to sign and acknowledge they have read and understood the safety rulebook.

Your safety rulebook can be in any format you want. The important point is that you cover as many areas as possible and empathize the fact the employee is responsible and accountable just like the company. Good luck.

Chapter III

"What's broken?"

Identifying the Losses to Your Company

Most companies use an insurance carrier as a defense against economic loss. But an insurance company tries never, never, never to spend its own money on your company's losses. There are no free lunches with the insurance industry. Yes, your company does pay the premiums, and yes, the insurance company does act as a bank, making payment when losses occur. But it is still your company's money. This is very important for the Safety Director and top management to understand. Insurance companies are in business to make money out of your premiums and if they don't, usually only two things can happen: 1.) They raise premiums or 2.) They say good-bye to you. It's that simple. Remember that insurance companies are not the villains they are sometimes made out to be. The vast majority do a good job providing you with protection against **your** losses.

Insurance Loss Runs

"Loss run," "claim report," or any other similar term simply refers to a tool designed to tell you how your company dollars are being spent by the insurance carrier. The vast majority of insurance companies issue some type of loss run or can provide you with claim loss information.

You need to have copies of your company's insurance loss runs. Your company controller may already be receiving them but not sharing the information with you. If so, you can call your insurance agent or your insurance company loss control representative to receive the loss runs. It is not enough that your agent "will handle it for you" or that your controller gets them but doesn't want to share. The Safety Director must have access to the loss runs.

Once you begin receiving the loss runs, you must review them.

Examining a Loss Run

The Insurance Categories

Loss runs look more intimidating than they really are. They usually are organized into categories by type of insurance:

General Liability to Include Operations and Premises:
These are claims made by people outside of your business who have had an accident/incident.

Property:
This is coverage for fire, wind, hail, flood, etc. It does not include auto.

Workers' Compensation:
This is insurance for your employees in the event they are injured in and out of the course of their employment. In most instances this is where the dollars go. (That means this is the coverage that loss prevention can have the greatest impact on.)

Auto:

This category applies to companies that have "auto fleet exposure" because employees use company-owned vehicles as part of their job.

Liability Product:

This is coverage for lawsuits or claims by customers or end users who are injured by a product that you manufacture.

Nine Injury Categories

Learn the nine basic categories of injuries:

Struck By	*Fall/Slip/Trip*
Body Mechanics	*Caught-In-Between*
Laceration/Cut/Tear/Puncture	*Eye*
Struck Against	*Miscellaneous*
Contact with Temperature Extremes	

You will be using these categories not only for OSHA log review, but also for loss run analysis and for your monthly report card, which are discussed later in this book. If these are new terms to you, don't be concerned. The categories are simply ways of reporting what happened to a particular individual.

Struck By:

The injured employee was struck by a vehicle, tool, or another employee who was using a vehicle or tool in the performance of a job.

Body Mechanics:

This is an injury resulting from use of the torso or limbs not caused by an external source. It includes strains, back injury or a cumulative trauma such as carpal tunnel syndrome.

Laceration/Cut/Tear/Puncture:

This category includes injuries by laceration, cut, tear, or puncture. The employee could have been hurt while using a tool or when brushing against the sharp edge of something. An injury as benign as a paper cut is classified in this area.

Struck Against:

Unlike "struck by," which is caused by an external source, struck against indicates the injury occurred when the employee hit something.

Contact with Temperature Extremes:

This injury could be a burn or frostbite.

Fall/Slip/Trip:

The employee lost his or her balance and was injured as a result.

Caught-In-Between:

This injury occurs when fingers, hands or arms are caught by in-running nip points or pinch points in equipment. An injury in which an employee's fingers are slammed in a car door would fit this category.

Eye:

This category includes any injury involving the eye.

Miscellaneous:

This is the category for any injury not otherwise classified.

Deciphering Your Loss Run

What are the coverages costing you? In your loss runs there will be dollar amounts showing what the claims activity has cost and what the reserves are.

An example follows on the next page.

Workers' Compensation Date: 1-1-97, 1-1-98 (policy period)				
Occurrence	*Medical Indemnity*	*Misc.*	*Expense*	*Total*
PD (paid)	$ 434	$ 658	$ 380	$1,472
OS (Outstanding)	6,283	4,600	1,200	12,083
TOT (total)	6,717	5,258	1,580	13,555

Let's look at the illustration as an example of how a loss report reads.

Workers' Compensation is the type of coverage you have with this insurance company. In this case, it is the workers' compensation insurance for your employees when they are injured or claim to be injured in and out of the course of their employment.

Date shows the policy period from January 1, 1997 to January 1, 1998.

Occurrence is a single event relating to a claim.

Medical is the money paid by the insurance company for medical care claimed by the employee as a result of a work-related injury.

Indemnity is the money the insurance company has paid your employee for lost wages due to disabilities, rehabilitation or death.

Miscellaneous includes expenses such as expert witnesses, police reports or surveillance of employees or workers' compensation cases.

Total simply shows the total amount for each of the expenses categories.

Paid. Means the money that has actually been paid out by the insurance company.

> *Outstanding* is the reserve that the insurance company has kept back for the employee. This is money they have not spent, but are keeping just in case they might have to use it.

Now, let's look at this case and see what the loss run tells you about what is really going on and what it is costing you.

The insurance company has actually paid $434 for medical treatment for the employee and has reserved $6,283. They have actually paid out $658 for indemnity (lost wages) and are reserving $4,600 of your money. For miscellaneous expenses they have paid $380 and are reserving $1,200.

The injury actually cost the insurance company $1,472. They are keeping $12,083 in the bank "just in case." However, it is costing your company $13,555 for insurance renewal. The insurance company counts the reserve as actual money paid out when they figure dollars lost at the end of the year.

Insurance companies are in business to make money, not to lose it—just like in your business! This workers' compensation case has actually cost the insurance company $.11 out of the $.89 they have charged you (see below).

$$\$12,083 \ / \ 13,555 = \$.89$$

When you start seeing the high reserves of the workers' compensation system, you will probably want to start screaming about how unfair it all seems to be. Yes, there are going to be times when no matter what you do you aren't going to prevent an injury or you will wonder if the injury is really fraudulent. Remember that it is your role to prevent injuries and be as professional about the situation as possible.

An insurance company loss run also breaks down the claims by incident, individual, date and type of occurrence. This breakdown is very important because it tells you what, where, when, how and why!

Do not leave the analysis of loss runs to the company controller or insurance agent. If you are going to develop and implement your company's safety program, you need to know your way around a loss run. Every insurance company's loss run is just a little different. Whatever yours look like, you must learn how to read and interpret them. Your insurance agent and/or controller should be able to go through one with you. Also, you can ask your insurance company's safety representative or claims representative for help. The Safety Director must know where the company's dollars are going!

Examining an OSHA Log

Many people who hold Safety Directors position keep excellent OSHA log information but never really use it. The OSHA 200 log is one of the best road maps you possess to find out the who, what, where, when and why of an employee injury or illness—don't waste this vital information.

The OSHA log is broken down by date, employee's name, occupation (job title), department and description of injury. Those five categories can lead you to better loss prevention decisions.

Bureau of Labor Statistics
Log and Summary of Occupational
Injuries and Illness

Let's use the sample OSHA log to determine what has happened so far in this company. Make two sets of headings on a scratch pad.

This is the first heading:

Date	*Name*	*Occupation*	*Dept.*

The second heading looks like this:

<u>*Dept.*</u>   <u>*Desc. of injury*</u>   <u>*# of Injuries*</u>

Once you have made your headings:

1. Total all of the injuries for January, February, March, and April.
2. List the last name of each injured employee.
3. List the job title of each injured employee.
4. List the departments where the injuries occurred.

5. Tally the types of injuries sustained by each department:

<u>*Date*</u>	<u>*Name*</u>	<u>*Occupation*</u>	<u>*Dept.*</u>
Jan. 4	Jones	Machine Operator	5
Feb. 2	Williams	Machine Operator	5
March 2	Anderson	Machine Operator	5
April 1	Wilson	Machine Operator	5
April 4	Smith	Lift Operator	4
April 9	Hawkins	Pallet Operator	2
April 12	White	Painter	3
April 22	Hall	Runner	1
April 23	Johnson	Machine Operator	1

OSHA Form 200

Note: This form is required by Public Law 91-596 and must be kept in the establishment for 5 years. Failure to maintain and post can result in the issuance of citations and assessment of penalties.

Recordable Cases: You are required to record information about every occupational death; every nonfatal illness; and those nonfatal occupational injuries which involve one or more of the following: loss of consciousness, restriction of work or motion, transfer to another job, or medical treatment (other than first aid).

Case or file No.	Date of Injury	Employee's Name	Occupation	Department	Description of Injury or Illness
1-1	1/4	Jones, John	Machine Operator	#5	Strained Back
2-1	2/2	Williams, will	Machine Operator	#5	Sore Lower Back
3-1	3/2	Anderson, Al	Machine Operator	#5	Sore Back
4-1	4/1	Wilson, Walt	Machine Operator	#5	Lower back Ache
4-2	4/4	Smith, Sam	Lift Operator	#4	Sore Lower Back
4-3	4/9	Hawkins, Hank	Pallet Operator	#2	Object in Eye
4-4	4/12	White, William	Painter	#3	Back Ache
4-5	4/22	Hall, Henry	Runner	#1	Back Ache
4-6	Johnson, Jan	Machine Operator	Machine Operator	#1	Back Ache

OSHA No. 200

Dept.	Desc. of injury	# of Injuries
1	Body Mechanic	2
2	Eye	1
3	Body Mechanic	1
4	Body Mechanic	1
5	Body Mechanic	4

With the situation laid out before you, you can find the common denominators.

Let's see what is going on in the plant represented by the example data:

Date: April accounted for six of the nine injuries. Make some hypotheses about the cause.

(Ask yourself: Was the plant recently re-opened? Were new employees hired or experienced ones transferred from a familiar job to a new one? Was a new process or new equipment installed?)

Name: Look for repeat names. In our hypothetical list there are no repeaters. If you do have the same name coming up you should look for training, attitude, supervision or personal problems, or mental/physical inability to perform the job task. Something is causing the problem.

Occupation: Notice patterns in job titles. In the example the majority of injured employees are machine operators from department five. Don't jump to conclusions about the significance of the pattern—a majority of your workers might be machine operators.

(Ask yourself: What type of work correlates with injuries?)

Department: This category tells you where in the company the injuries are occurring. In the list, Department #5 has four (44%) of the injuries, followed by Department #1 with two (22%).

(Ask yourself: Where are the injuries happening? Who is the supervisor in each case?)

Description of Injury: This category tells you what type of injury occurred. Accidents can't be prevented if you don't know the type of injury. On our list, eight of the nine injuries are body mechanic injuries (89%). "Body mechanics" include strains, sprains, back injuries and any type of injury that movement of the body can cause.

(Ask yourself: Why are these kinds of injuries happening? Make a note to yourself to learn whether supervisors and employees have been given body mechanics or lifting training.)

Let's now accumulate our data and see what the OSHA log has given us to find common denominators.

1.   April has been the worst month so far.
2.   There are no repeat names in the list of injured.
3.   Machine operators have had the most injuries.
4.   Department #5 has accounted for nearly half of all recordable injuries.
5.   Body mechanics injuries have accounted for 89% of all recordable injuries.

Take a moment to go through these facts to determine what is relevant and what is not.

Okay, the first thing to note is that your employees are lifting wrong or using their bodies incorrectly in some other way. The injuries could be caused by an ergonomic problem ("ergonomics" means fitting the job to the worker rather than forcing the worker's body to fit the job), or the reason could be the lack of lifting training. The thing to note is that body mechanics is what is wrong. Look at your insurance loss runs to determine what a body mechanic injury is costing you.

Department #5 is where your problems appear to be. Once you know this, you need to ask further questions. Is there a legitimate reason why one department accounts for 44 percent of all recordable injuries only four months into the year? If there is no logical reason, could this large number of injuries be caused by a lack of training or lack of supervision or both?

Whatever the reason for these injuries, your OSHA log tells you that there is a situation that needs to be investigated and corrected.

Addressing the Trouble Areas

This manual was not written to tell you how to correct what is broken, but rather to show you how to discover what is broken and then give you the tools to start the repair process. This process will include those people who have the wherewithal to assist you. Again, these people will include your insurance company loss control representative, claim rep., OSHA, and safety supply representatives. Also, your training from the NSC will address various safety topics including personal protective equipment, materials handling and storage. With this information you can then begin to fix "what's broken."

The Real World: Identifying the Losses to Your Company

Bumps in the Road

It is possible for Safety Directors to have unlimited power. They are the ones who know, or can find out, what the law says your company must do. This is usually very threatening to those department heads in your company that are entrenched. They see the new Safety Director as someone with a newly created position who presumes to believe that he or she is going to be an equal. That goes down hard for some department heads, so be aware of it.

Controllers and the Safety Director

Nothing cripples your efforts to find out where your company's money is going more than the denial of access to your insurance company's loss runs. But if you are not in the position to walk into the controller's office and simply state that you need access to them, what then?

The company controller may not want to share them, or may block you for no other reason than he or she simply doesn't think this is material for you to know. Many times I have found that the controller already knows where the dollars are going but won't rock the boat to get action to stop the problem.

Take the problem to your boss. Simply say, *"In order for me to do my job I must have access to the loss information so I can show where the losses are occurring and what the average accident is costing us."*

In some companies a controller will furnish you with a recap of the loss runs and that's all. If a recap is all you can get and if that is fine with your boss, then you may have to learn to live with it

Entrenchment Blocks the Safety Director

Here's an example of blocking by entrenched personnel. It occurred in a large electronics corporation that was one of my accounts when I worked as a Loss Control Consultant. They were having serious workers' compensation problems and knew that something had to be done. The personnel manager was told to find someone in the department to start a program to get the insurance company (me) off their backs.

I met with the Safety Director, a young female assistant personnel manager. She had done her homework. By using her OSHA logs and by getting the loss runs directly from me, she and I had pinpointed where the problem was. Now it was up to the two of us to solve the problem.

When I met with the safety committee for the first time, the chairs in the room were filled, but the assistant personnel manager was not there. I asked the chairman of the committee why she was missing. Nodding and smiling, her boss told me he would be representing her. His young assistant need not attend a meeting of this importance.

I addressed the room of executives, saying there were only two people who knew what the problem was and how it should be fixed. I emphasized that their company employed only one of them, and that she wasn't being allowed to attend.

Upon my insistence, the woman was brought into the meeting. Ironically, she was not asked to sit at the conference table, but rather behind her boss. I conducted the meeting and called on her to brief the big wheels as to the problem and solutions we were going to take.

After three similar monthly meetings, she was finally allowed to sit at the conference table. It always amazes me how companies will ask someone to do a job, yet put up barricades to stop him or her from being successful. That, however, is the real world. In this case, the young personnel assistant knew what needed to be done, and because of that knowledge the company was forced to listen to her.

Chapter IV

"How bad are we?"

Comparing Your Company to Others

How does your company stack up against others in your industry? Many times I find that a new Safety Director not only doesn't know what is broken, but also is absolutely convinced that the company has a much worse safety record than the national average. So how do you find out how you are really doing?

Performance Measurements

An Essential Resource

If you don't already have the National Safety Council's *Accident Facts* book, buy it! It is inexpensive and will give you much valuable information. Simply call your local safety council or the National Safety Council and they will tell you how to obtain a copy.

Next you will need to know what your company's *Standard Industrial Classification (SIC) code* is. You can get that information from your insurance agent or controller. If you now have access to the Accident

Facts book, turn to the section on workers' compensation, which lists all the various types of industries by name and SIC code. (If you don't have the *Accident Facts* book, please follow along anyway.) For example, let's say you are in the fabricated metal products business. Run your finger down the left-hand column until you find fabricated metal products and then across to the SIC code. It tells you that you are **SIC code 34**. At the top of the page it has *"total cases," "lost work day cases," "non-fatal cases without lost work days,"* and *"lost work days."*

Right now the procedure seems a little overwhelming, but it's not that difficult. At least you now know what the average is in your business, so let's find out how you compare.

Calculating Your Incident Rate

The *incident rate* is figured by multiplying the yearly total number of injuries and/or illnesses by 200,000. The number 200,000 is an arbitrary number used by the safety industry to represent one hundred full-time workers working forty hours per week, divided by total workers' hours for fifty weeks a year. It is a simple measuring device that any industry can use to compare itself against its peers. Let's try this formula.

Assume it is now the end of the year and your OSHA log is complete.

1. You have **20** recordable injuries for the entire year.
2. You had **4** lost time injuries.
3. You had **486** days lost.
4. Your employees worked **566,540** hours.

Your incident rate is calculated as:

$$\frac{(20 \times 200{,}000)}{(566{,}540)} = 7.06$$

The incident rate is 7.06. What does this figure mean? It means that for every one hundred of your company's workers in the previous year, 7.06 of them would have had a recordable injury or illness. It makes no difference whether you have ten employees or 10,000 employees—the incident rate is an index that represents how you are doing.

Now run your finger across the column in the SIC code 34 row to total cases. It shows that the national average for total cases is 17.0. Remember that your company is 7.06, so the incident rate comparison reveals that you are significantly lower than the national average!

To see how you compare on the number of *lost time injury cases*, you multiply the number of lost time cases (4, as previously stated) by 200,000 and again divide by the number of hours worked (566,540).

$$\frac{(4 \times 200,000)}{(566,540)} = 1.41$$

Your lost time injury cases index is 1.41. When you run your fingers across the page, you will find that the national average is 7.2. Again, you are much lower than the national average!

Now let's compare your company with the national average for *lost work days*. With this figure, you can compare your company with the national average for number of days away from work that injured employees take. As identified above, you had 486 days lost, and your employees worked 566,540 hours.

$$\frac{(486 \times 200,000)}{(566,540)} = 171.5$$

Your lost work days index is 171.5. When you run your finger across the page, you see that the national average for your SIC code 34 is 121.9; yet the number of days your employees take off is 171.5. Here you are higher and alarm bells should be going off in your head.

When employees are off work on compensation, you need to mentally picture a taxi meter clicking off the dollars that your company is paying. Don't picture the insurance company paying, because it is your company paying!

Putting It All Together

Now let's take a look at the big picture. Here's your procedure for finding "what's broken?"—or for that matter whether things are as bad as you thought.

1.   Insurance Loss Runs (covered in Chapter 3)

    a.   Get insurance loss runs from your agent, company controller or insurance company loss control representative.

    b.   Have one of these individuals take the time to show you step by step how to read and interpret the figures.

    c.   Once a month, examine your loss runs and record your observations. (You need to know the current loss run information, so analyze it at least quarterly if not monthly.)

Don't make the mistake of letting your agent or controller simply advise you of your losses. You need to know where the dollars are going. (If I were your boss and asked you, the Safety Director, what our losses were costing us and was told that the controller has that information, I would have very little confidence in your knowledge or ability.) Remember that the old adage is correct—"Knowledge is power." In the safety profession you will find that you are going to need all the power you can muster.

2.   The OSHA Log

    a.   Use the OSHA log at least once a month to look for patterns and trends.

    b.   The log can give you the what, where, who, how and why.

Remember: your boss is going to be much more impressed with you when you can tell him or her exactly what is going on. Use the log as a persuasion tool!

3. The Incident Rates

a. Find out how you measure up. Use the incident rates to determine where you currently stand in comparison to others in your industry.

b. Practice calculating incident rates for your company until you can be comfortable with both the formula and using the accident facts book.

Let your boss know how your company compares to your industry rates by showing your calculations and comparisons to him or her.

Now that you know what is wrong, let's move on to the process for how you are going to start down the road to correct it.

The Real World: Comparing Your Company to Others

Explaining the Incident Rate

Comparing your company to others by using the OSHA incident rate and national accident facts information *should* be an excellent tool to show others how you really do compare. Department heads and other senior staff usually understand, *but it is critical that you fully educate the first line supervisors on the concept and calculation of the incident rate.*

Try to determine the knowledge level of your supervisors and gear your education of the incident rate to that level. Be descriptive by comparing the incident rate to report cards, performance evaluations, or other common measuring tools. Drive home examples of how it honestly reflects what is going on in your company and in the respective departments. Statistics and formulas are threatening to most people, but if you can break the incident rate concept down and make it real by showing how it works as a measuring device, chances are that the supervisors will be more responsive.

The Unfairness of It All

Your first line supervisors may accept the incident rate as a measuring device for their departments when you initiate the concept. However, when their performance evaluations, pay raises, and/or bonuses start to be affected by it, they may complain that it is unfair. You will hear time and again that there is no way to prevent a fraudulent workers' compensation claim or a machine malfunction that resulted in an injury to one of their employees. There are times when they are right: some occurrences are beyond their control and that really is unfair. But ninety percent of all injuries to employees are the result of unsafe acts by the employee. And it is the supervisor who is responsible for the employee.

I have been tempted to make exceptions to the rule for the incident rate, but I never did. I always require that the injury be recorded by the department that it occurred in. If you start making exceptions for certain cases, the whole intent of the incident rate as a measuring tool will be lost. Stand firm on this. Some supervisors will attempt to manipulate, cajole, threaten and whine their way out of accepting their responsibility to their employees, but don't let them. The first line supervisor is the key to the success of your program.

Chapter V

"That sounds great, but..."

Selling Your Boss on the Program

Explaining to you the need and purpose for the Loss Run, Incident Rate, and OSHA Log are fine in that it gives you a base, a starting point for your program. Regrettably, it often does not impress your boss one little bit! One of the biggest reasons it does not is because Safety Directors are usually very poor in breaking down the need for support of their program based on money. It often really is that simple. Okay, let's talk about money.

Safety programs are not in place in American Industry because one morning the Boss woke up and had a revelation that something was needed to protect the workers on the plant floor. CEO's and Plant Managers support safety programs for two reasons: They get better worker compensation rates and fewer OSHA citations if they do.

Granted, there are companies, albeit very few, that are motivated to protect their workers because it is indeed the right thing to do. But, in the vast majority of cases it is predicated on money, the bottom line. Safety is almost always about money. Often Safety Directors fail to grasp that concept and flounder in their persuasion of their boss in spending the funds needed to support their program.

Where Are the Dollars Going? The Loss Run

Your boss, if he or she is like most bosses, is driven by money. It is always the bottom line in business. Can you go into their office and show them where the dollars are going from preventable injuries? In the previous chapter we detailed how you were to decipher an insurance company loss run. But, what are you going to do with the information? How will you sell the information to your boss so they will not only listen to you, but also give you the support you need? Safety Directors in many respects are often the world's worst marketing people. They feel that their cause is noble and virtuous and they either should not have to sell it, or they lack the skills in selling it. Well, Safety Director, you simply have to if you want to succeed.

Here are some questions you should ask yourself about the Loss Run:

1. What type of injury is presently costing your company the most in expenses? Go back to your nine injury categories and make a list of where your insurance dollars are going for what type of injury.

2. What type of injury has the highest Medical expense? Remember Medical? That's the money your insurance carrier is paying for actual medical care. What type of injury is driving this expense? Why? What is it that can be done to prevent this cost?

3. What injury has the highest Indemnity cost? Indemnity is, again, the money that your worker's compensation carrier is paying your employees for lost wages due to disability and rehabilitation. It won't take you long to realize that Indemnity is the cost driver in your loss run. Why is one type of injury more expensive than another? Usually it is body mechanic injuries. Those dreaded strains, sprains, backaches, and the feared Cumulative Trauma Disorders. Know your facts on indemnity! Be able to nail the corners down on where the money goes, as it almost always tracks back to this expense.

4. When reviewing the above three items be sure to know what has been actually paid. Paid is again, the money your insurance carrier has paid, has written a check and issued on this particular injury. Check each category for money actually paid out.

5. Sort out the reserves the insurance carrier has held back, their fudge money to make sure that they will have enough funds to continue to pay the bills in the event the injured worker needs more medical care or staying at home time. Watch this expense; remember that you're being charged for this money even though your carrier has not yet spent the money. At renewal time, if the claim is still open, your premiums will reflect their outstanding/reserves on the injury. Remember to include all the categories from your loss run. Okay, you have broken down the money from the loss run and you have your Figures. Let's now go to the next leg in your stool: The OSHA Log.

The OSHA 200 Log—Your Injury Roadmap

1. What department is having the most injuries by type? List this for your research.

2. What occupation is having the most injuries by type?

3. What type of injury is occurring most often?

4. What individual is having the most injuries? You need to know who is costing you the most money? Safety is a people business.

Many Safety Directors like to hide in the lofty world of facts and statistics. Although this posture is comfortable, it will not resolve unsafe acts and unsafe conditions created by people. It is your job, Safety Director, to determine the "who" part of your research. Let's now finish up with your Incident Rate.

Your Company's Incident Rate—Your Safety Process Measuring Device

1. First, how does your company compare with other "like" companies in your overall incident rate? Be cautious here. You may be about average or even below the national average. That in itself, however, does not necessarily represent where you money is going. Write down your incident rate and move on to the next category, Lost Time Injury Cases.

2. Lost Time Injury Cases is where the cash register really begins to ring in your lost injury dollars. Lost time injuries involved the indemnity part of your insurance. This is the money paid to your employees to stay home and recover. These lost days drives your insurance premiums up and up. In the eyes of the insurance carrier you have injured the employee so severely that he or she simply cannot return to work. In a later chapter we will discuss the return-to-work program. Your lost time injury cases are a critical part of your research to substantiate your program to your boss. What could be worse than Lost Time Injury Cases? How about the number of Lost Days?

3. Lost Time Injury Cases reflect how many injuries have the potential to cost your company big bucks. Lost days is where the numbers begin to add up. It cost you money when you have a recordable injury. It costs your company more money when the employee cannot come back to work because of the injury. The number of days away from work, Lost Days, is your focus. Compare your Lost Days Incident Rate to other like companies. If yours is higher then you have an expensive problem. It's bad enough to allow employees to be injured. It's even worse to injure them so bad that they cannot return-to-work after immediate medical care. The worst possible scenario, other than a massive injury or death, is long term rehabilitation of this injured employee.

Now What?

Let's break down what you have so far by giving you an example of how your research should reflect what is going on. I'll use a fictious example.

Loss Run

1. Your loss run research has shown that Body Mechanic injuries are occurring most often.
2. Body Mechanic injuries also account for the most medical expense in your loss run.
3. Also, Body Mechanic injuries are responsible for the highest indemnity cost.
4. Again, Body Mechanic injuries have the highest insurance dollar reserves.

Conclusion: Body Mechanic injuries, strains, sprains, Cumulative Trauma Disorders are costing your company through insurance expense, thus premiums, the most money. Body Mechanic injuries equal high dollar loss.

OSHA 200 Log

1. The warehouse department has the most recordable injuries.
2. The occupation of warehouse worker has the most recordable injuries.
3. Body mechanic injuries are the most frequent, with back strains and sprains. These injuries are occurring most often in the warehouse.
4. Sam Smith, Warehouse Worker has had the majority of the body mechanic injuries. To date he has accounted for 37% of all dollars spent on body mechanic injures.

Conclusion: Body Mechanic injuries occurring in the warehouse department by warehouse workers are costing your company the most money. So far you know what is costing the most in your insurance dollars, and know you know where the injuries are occurring.

Incident Rate

1. Your overall incident rate is 25.7 compared to the national average of 16.4. This reflects that you are 36% above the national average. Your company has 36% more recordable injuries than a similar company doing the same work with the same exposure.
2. Your lost time injury cases are 12.3 compared to the national average of 5.5. Your company has 55% more lost time cases than the national average compared to a similar company. Remember that lost time cases start the cash register ringing.

Conclusion: Not only are you injuring your employees at a higher frequency, but the severity of the injuries is also worse.

3. Your lost days are 122 compared to the national average of 67. Again, your company is not doing as well as other like companies. Your company has 45% more days at home required for employees to recuperate from their injuries.

Conclusion: Not only are you injuring your employees at a higher frequency and severity it is also taking your employees longer to heal.

With this information you can go to your boss and talk money. Injuries equal dollar loss. Develop in your own method a system of addressing the dollar loss to your company. In most instances this is when your boss will listen to your needs. Talk bottom line dollars. Your boss may mouth support for your Safety program, but it almost always is the financial expense that will get his attention.

Date:
To: Your Boss
From: You
Subject: Injury financial expense

I have conducted a financial analysis of our injury expense by using our insurance company loss run, OSHA 200 log, and our incident rate in comparisons to other like companies.

1. Our insurance loss run reveals the following:

   A. Body Mechanic injuries are occurring most often on our loss run.
   B. Body Mechanic injuries are on average our most medical expensive injury.
   C. Body Mechanic injuries have the highest insurance indemnity expense.

2. Our OSHA 200 Log reveals the following:

   A. The Warehouse Department is having the most injuries.
   B. The occupation of Warehouse Worker is the most frequently injured worker.
   C. Body Mechanic injuries are the most frequent injury in our warehouse department.
   D. Warehouse worker, Sam Smith, has had 4 recordable injuries resulting in 37% of the total dollar loss from Body Mechanic injuries. I have contacted his supervisor regarding Mr. Smith.

3. Our Incident Rate reveals the following:

   A. Our overall recordable injury incident rate reveals that we are 36% higher than the national average.
   B. Our Lost Time Injuries are 55% higher than the national average.
   C. The number of our lost days—those days our employees need for recuperation from their injuries—are 45% higher than the national average.

   I would like to meet with you to discuss options I have developed to reduce these injuries and subsequent expenses.

It is not enough to be able to research the material. You must be prepared to meet with your boss with solutions to your injury problems. Otherwise you will probably hear him or her say, "That sounds great, but..."

Chapter VI

"It's not my job."

Safety and the Supervisor

In the world of safety, no one subject has been written about as much as the first line supervisor's responsibility in regards to preventing injuries.

Why Involve the First Line Supervisor?

Your first line supervisor have control over your employees in the areas of quality and production. They should be using that same control in safety. Because supervisors are the eyes and ears of the company for quality and production, the same is true for safety.

Control

Your first line supervisor have control over your employees. No one other than spouses, relatives or close friends has that type of influence. At work, these supervisors are the ones who make the determination as to what your employees' day is going to be like. They determine the work schedule and the quality and quantity of work that your employees will produce. Your employees, whether they admit it or not, watch supervisors to learn the importance and significance of a company policy. What you are trying to accomplish is even dependent upon your supervisor's facial expression, body language and general attitude.

Don't kid yourself into believing that just because your supervisors mouth their acceptance of your efforts, they will enforce them on the plant floor. A roll of the eyes or a shrug of the shoulders can do as much to undermine what you have put together as an open statement saying they think safety is a waste of time. Conversely, a supervisor who honestly believes that a safe plant is the only way to go can win extra mileage for your safety program that you will never achieve on your own. Therefore, it is so very important to train, train, and train supervisors. With knowledge comes power. Most people, and especially supervisors, like power.

Your Eyes and Ears for Safety

Who else on the plant floor knows as much as your supervisor? Who else really knows whether or not an employee is following your safety policies and observing safety rules and regulations?

A walk-through by top management is important, but where safety is concerned, nothing takes the place of the minute by minute, hour by hour, day by day supervision of your first line supervisor. When supervisors are trained in safety—and they must be—they are your eyes and ears, your enforcers and counselors to get the job done, to save your company money and to comply with federal and state regulations. You need to cultivate supervisors and reward them for their efforts.

Supervisors and the R.A.A.C.K.

The first line supervisor must be reached. You as the Safety Director need to develop a system of motivation and accountability. In all my years in the safety business, one of the toughest things to accomplish was to get the supervisor on the R.A.A.C.K. The R.A.A.C.K. is my acronym for responsibility, authority, accountability, consequences and knowledge. Let's look at these five areas:

Responsibility—First line supervisors must answer to upper management for their activities. They are responsible for the safety of their employees.

Authority—Here the first line supervisor have the right to make the necessary decisions to keep their employees working safely and to maintain a safe work environment.

Accountability—The first line supervisors' responsibility and authority are measured by upper management to determine if supervisors are indeed controlling accidents/incidents in their departments.

Consequences—Like the line employees, the supervisors must understand that there are consequences for involvement in a safety program and enforcement of it. Negative consequences range from a verbal warning, written warning, suspension and finally termination. Positive consequences can range from a simple thank you—to a better yearly evaluation, impact on the annual pay raise, or a bigger bonus. The supervisors must be made to know that no matter what kind of cooperation they make, there will be consistent consequences.

Knowledge—No one was born knowing anything except hunger and thirst. If you want your supervisors to push your program with their employees, then you need to give them knowledge about safety that they will in turn give to their employees.

Tools for Knowledge

Let's begin with the last of the items in the R.A.A.C.K., "knowledge." Don't think for a moment that you can have responsibility, authority, accountability, and consequences without giving your supervisors the required safety knowledge.

When dealing with knowledge for your supervisors, keep in mind the acronym "KISS," which means, "Keep It Simple, Stupid." I am not implying your supervisors cannot comprehend safety training, but rather, to educate anyone, information needs to be presented in a way that is very understandable.

I recommend the training that I used first with my supervisors: the National Safety Council's *Supervisor's Development Program*. The National Safety Council understands the need for supervisor training and the need for "KISS." They deliver the information in a positive, interesting, and entertaining manner. There are fifteen different training modules ranging from safety management and accident investigations to machine safeguarding and ergonomics. When I started my safety program, I was well aware of the National Safety Council's program, but I still researched other firms for comparison. The National Safety Council's *Supervisor's Development Program* is still far and away better than other programs. The videos, textbook, and workbook give your supervisors everything needed to teach them the safety knowledge they must have. Contact your local or state safety council or the National Safety Council for information on how to obtain the course.

This program will teach you how to teach your supervisor. It will hold your hand and take you step by step to the fulfillment of required training for your supervisors. The instructor's kit that you will be using is just as important as the student materials. It comes with a complete set of teaching tools to increase your understanding of the program and gives you everything needed for you to teach the program. Another plus to the *Supervisor's Development Program* is that it will fit your business and is adaptable. If you can't teach at work there is a home study course.

Don't worry about how to set up the classroom. I have suggested the National Safety Council's program because the instructor's manual breaks it down into a step by step sequence to follow: the classroom setup, the equipment you will need such as chalkboard, projector or tape recorder, and VCR for training tapes. It covers the preparation you will need for the topic and also includes the lesson plan itself, guides for the discussion, and suggested questions to keep on course.

No matter how easy the instructor's manual makes it sound, the idea of teaching can be intimidating. I suggest that you get copies of the manual and textbook and take yourself through the course. You will be surprised at how it all falls into place for you. Highlight areas in the text, manual, and visual aids to make special note of items that are relevant to your company operations. After you take the course yourself, you will have much more confidence. And your newfound expertise will help you to start to fix the "what's broken" that you discovered in your OSHA log and loss runs.

Tools for Accountability

Your supervisor accountability program assumes that positive evaluations of a department are the result of good safety work by the supervisor.

Department supervisors need to see how they are doing in comparison with their peers. The incident rate formula is a good tool for that purpose. In my job as a Safety Director, I ran a "monthly cumulative incident rate" for each department. I met with the department supervisors and explained the system to them. Using the forms, they could then see how they were doing month by month in comparison to the national average as well as in relation to their peers. Also, I mounted a graph board in my office which showed visually how they were doing—it received a lot of attention! As the months went by and the number of employee hours worked went up, their incident rate went down.

Also included in the Appendix is the monthly breakdown of dollar cost by each department. I used this tool to show the department supervisors what the injuries in their departments were costing the company as a whole. This information came from their insurance company loss runs (another reason why it is so important for you to see the loss runs).

The supervisor's yearly evaluation is another tool for accountability. The nine criteria for evaluating supervisors break down the activities and involvement required by the first line supervisor. Show the first line supervisors how they are doing month by month and let them know that their efforts or lack of effort will be held accountable. (Copies of these forms can be found in the Appendix.)

Tools for Consequences

After you have measured what the supervisor has accomplished, there needs to be consequences for these actions. A successful company sits on the three-legged stool of *quality, production, and safety*. Safety is the newest leg, and you will find it difficult at times to get the supervisors as committed to the safety program as they are to quality and production.

If a supervisor has done an honest and coactive job in assisting you in reducing injuries, he or she should be acknowledged in some way for that effort. If, on the other hand, the supervisor has not gotten on board, the consequences must be serious enough to get this person's attention. Don't back down from what needs to be done with the non-compliant supervisor. *If you back down, your program will fail!*

Negative Consequences

Negative consequences for the supervisor are the same as those you will use for the line employee: 1.) *verbal warning*, 2.) *written warning*, 3.) *suspension*, 4.) *termination*.

The supervisor must be held to a higher standard than the line employee. Remember, your first line supervisor is the key to the success of your program. There should be no doubt in the supervisor's mind that he or she must enforce the program or risk termination of employment by not doing so! At first, disciplining for poor adherence to safety policy may appear heavy-handed and unnecessary, but think about the consequences in your company for a supervisor who allows shoddy work to come off the line, or a supervisor who does not maintain order in the department. What are the consequences for these kinds of incompetence? Shouldn't safety be treated as seriously?

Positive Consequences

Positive consequences are just as important, if not more important, than negative consequences. Personal incentives include higher evaluations, bigger pay raises, and better bonuses. But you can also work on the department level. Departments that do not have injuries that are clean and orderly and do not cost the company money need to be rewarded by word and deed. *Praise not spoken is praise not given.* A memo to a department supervisor thanking him for his efforts to clean up his department or for going a month without a recordable injury, with a copy going to the president of your company, will go a long way to prove to you the power of positive consequences.

I have used the concept of "Team Safety of the Month," a program that gave the supervisor recognition. To be Team Safety of the Month, a department had to meet three criteria:

1.   No recordable injuries for the month.
2.   All employees in the department follow safe work practices and use mandated personal protective equipment.
3.   Good general housekeeping.

As I would make my daily rounds, I would rate the departments on a scale from 1 to 5, with 1 being low and 5 being high. Supervisors and line employees knew what I was doing because I explained the system to them. They found it fair and equitable and made conscious efforts to improve their work areas.

The yearly evaluation of your supervisor's performance is crucial to the success of your program. Right now, take a look at the Safety Performance Evaluation (Exhibit S in the Appendix).

I developed the form to make it as simple as possible for me to fill out and for my supervisor to understand. The scale of 1 to 5 is easy to grasp and does not allow that much room for abstract gray areas of performance. In other words, the supervisor is either doing the job—or not doing the job—better or worse than his or her peers.

Keep On Track by Measuring

When dealing with supervisors it is important that you remember this: *That which is measured, is done.* Measuring the supervisor's position on the five parts of the R.A.A.C.K.—responsibility, authority, accountability, consequences and knowledge—will keep your supervisor on track.

Make no mistake about it, a safety program means *work* to the first line supervisors. It will not be fun for them and in many cases it will be strongly resented. They feel, with some justification, that they have all they can handle as it is, and they don't want or need additional responsibilities and work. "That which is measured, is done" means you must measure their activities, their effort, and their commitment When you calculate your monthly incident rate by department, how do supervisors compare with their peers, with the rest of your company, and with the national average? As you make your inspections, how orderly and safe are their departments and could they make them safer and more orderly? What is their commitment to reducing injuries and saving the company money?

Nine Criteria for Evaluating Supervisors[1]

Nine criteria are basic to a supervisor's activities:

1. *Does the supervisor ensure his employees' compliance with safety rules and regulations?* As you walk through the plant, are the employees following rules properly? Is the supervisor enforcing these rules and regulations? You need to keep a logbook on each department's supervisor and make notes on what you find.

2. *Does the supervisor condone unsafe behavior for the sake of production?* In my work I have discovered that employees who are high producers are allowed to cut corners on safety, whereas the average employees are held more accountable. Such supervisors know they are allowing this behavior, but more importantly, the other employees know the supervisor isn't playing fair. Their

---

[1] Supervisor Safety Performance Evaluation, Appendix S

perception that a supervisor puts production before safety will kill your safety efforts. Don't allow supervisors to cut corners on compliance and enforcement.

3. ***Does the supervisor use acquired safety knowledge?*** You have spent the company's money and time training your supervisors, and they now need to put that knowledge to use. Don't allow them to say they didn't know better when in fact they learned it in training. One technique I used when I found a supervisor not using newly acquired knowledge was to have him or her look it up in the textbook and bring me the answer. The word soon spread that supervisors were accountable for using the knowledge and resources they were given.

4. ***Do supervisors listen to their employees' ideas on ways to improve safety in their departments?*** Do they encourage employees to make suggestions? Can employees volunteer an idea without fear of rejection or silent dismissal? Your line employees are a valuable source of information to improve the safety and efficiency of the operation. The supervisor needs to listen and to facilitate the generation of new ideas.

5. ***Do supervisors solve safety problems actively, resourcefully and independently?*** Criterion five is one I used as a benchmark to see how successful my training with the supervisor was. It was also a very good measuring device to determine if the supervisor was motivated to make his or her work area safer. Your supervisors came equipped with minds and the ability for original thought. Are they following the map you have laid out for them, or do they come to you with every little problem so you have to make the decisions for them? Are they resolving departmental safety issues by using their own initiative to solve those problems?

In the beginning the supervisors will, of course, be hesitant and reluctant to decide on their own, and that is understandable. But when they begin to see what is needed, they should be able to make decisions based on their own resourcefulness and initiative. Remember, this item was put into the evaluation so you would not have to be the department safety manager—that is their job!

6. *Do supervisors really make their rounds?* Criterion six notes that your supervisors are equipped with eyes, ears, and legs. Supervisors need to be in their departments making rounds, talking to the employees, and observing their work habits regarding safety. If they see an employee violating a safety rule, are they making on-the-spot corrections? Are they instead choosing to wear what they perceive as a "white hat" and be the "good guy" who does not enforce the safety rules that are needed? A good way to determine if they are making rounds on a daily basis is for you to make rounds also. If, for instance, you find the same employee not wearing his safety glasses day after day, isn't it reasonable for you to assume that the supervisor is also seeing this, but not making corrections to the employee's behavior?

7. *Does the supervisor properly handle accident/incident investigation, and paperwork?* Your supervisor will be trained in accident investigation and needs to complete one every time there is a recordable injury. (See Chapter VII in which the accident investigation is discussed.)

   Is there depth to the report? Is it readable? Does it give a complete picture of what happened? Was it given to you as soon as possible after the injury with a good explanation of what needs to be done to prevent recurrence? If not, the supervisor needs to know that there are going to be consequences for failure to complete this important form. The evaluation is a measuring device for this activity.

8. *Has the supervisor improved the accident statistics for the department?* Once the supervisor is taught to understand how to calculate the department incident rate, as well as to understand your company's SIC code and the national average, you can move to using comparative statistics as a criterion for giving rewards. Has a supervisor reduced a department's incident rate? Rates are measurable; much emphasis can be placed on them. Reduced rates mean that a supervisor's employees are having fewer injuries and costing the company less money.

9. *Does the supervisor talk straight and play fair?* I have dealt with supervisors who had difficulty in grasping the concept of a safety

program and the need for it, but with good intentions did their level best to reduce injuries in their departments. They were not smooth or sophisticated in their approach, but they did make an honest effort to comply. In time these supervisors were successful, and they came to understand the need for a safety program.

Conversely, I have dealt with supervisors who took all the training, nodded and smiled at the right times, and yet when management (or the Safety Director) was not around would disregard the safety program at every opportunity.

The majority of your first line supervisors will support you if given straight, plain, logical reasons for the program. This item is in the evaluation for the purpose of measuring the efforts of supervisors who are as interested and motivated as you are.

Special Exemptions for Special Supervisors?

In every organization, I find there is a supervisor who is honest, hardworking, loyal, and dedicated, and yet whose employees are suffering a disproportionate amount of injuries. Get ready to confront your company's tolerance of anyone who is currently exempted with a "yes, but in this case...." It's hard to be the one to create severe consequences for such a likable and loyal employee. You will be told that top management wants losses reduced, wants the insurance premium to go down ... "but in this case the supervisor has been loyal and true for all these years and let's not really put the pressure on."

This is precisely the supervisor you want to create serious consequences for. Consistent treatment sends a strong message.

You and the company are totally committed to reducing accidents/incidents. If the supervisor is loyal, hardworking and dedicated, then getting on board shouldn't be so difficult. Don't allow "special exemptions"—supervisors either save the company money or they cost the company money, whether by action or inaction. If their behaviors and attitudes are costing the company money, there must be consequences!

The Real World: Safety and the Supervisor

Supervisors are people just like you and me. They want to be liked, just like you and me. Most people don't want to seem like a villain, so they don't want to enforce any more rules on their employees than is necessary. The Safety Director tells them they must do even more enforcing than they already do, and supervisors usually don't like that.

If you already know your first line supervisors, then you will know who doesn't have integrity and who does. You will know whom you can trust and whom you can't. You will know who talks straight and who plays fair.

When Production Is Behind

I can't think of any one subject that has as many false oaths sworn to support it than safety. No one is against safety—it's like being opposed to good mental health or Mother's Day. But when production is behind and people are being pushed, are your first line supervisors still enforcing rules and monitoring for unsafe acts and conditions, or is safety being ignored?

These supervisors don't work for you, but rather for their own department head. They may play you against that person to see who wins.

You must hold them and the department head accountable in whatever fashion your boss or your company will allow. Don't back down, not once. If your policy was important at the time you implemented it, it is just as important now, particularly when there is a production problem. There are going to be times that top management will overrule you, even if it violates written standards. If they do that, ask them to tell you specifically why it is being done—what is different now from the way it was a week ago?

Tell the department head, your boss, and the first line supervisor just exactly what OSHA standard your company will be violating, and emphasize the monetary consequence of violating that standard. It won't hurt to also tell them that they are sending a mixed message to the employee that will ultimately hurt the safety program.

Caught Between the Supervisor and Employee

Not every employee likes the supervisor. Employees find the Safety Director to be a great vehicle to get a supervisor into trouble with management.

Line employees must always be encouraged to tell you what they think about safety conditions in their departments and what needs to be fixed, such as through the Safety Suggestion/Hazard Report form (Appendix U). But you can easily run into a question of "who's telling the truth?"—the employee or the first line supervisor. An employee will tell you that a department supervisor doesn't care about safety and thinks it's a joke, but the employee doesn't want to be quoted for fear of retribution from that supervisor. Sometimes these allegations are true but sometimes they aren't. So what are you supposed to do?

The way I handle each and every case is to give the department supervisor the benefit of the doubt. I ask them directly about the incident without stating who told me. After a while I learn which supervisors and employees are playing fair and talking straight and which aren't. This course of action will not put you into the perceived role of a headhunter.

Chapter VII

"He did what?"

The Accident Investigation

The primary purpose of the accident investigation is to *prevent the recurrence of the accident or incident*. The investigation will attempt to prevent the same thing from happening again the same way. The accident investigation determines the what, where, when, why and how of an employee injury. Supervisors learn the basics of accident investigation and how to interview the injured employee and witnesses during the *Supervisor's Development Course*, or whatever other accident investigation training you choose to use. The accident investigation report is forwarded to your boss and to the first line supervisor's boss for their review.

The accident report is a major tool in your arsenal of weapons to prevent recurrence. Remember that your insurance carrier counts injuries/accidents in categories of frequency and also of severity. If you stop the frequency, the odds are in your favor that the chance of severity will be reduced.

Accident/incident investigation is not a legalized witch-hunt to find fault. *You are fact finding, not fault finding.* You review the accident/incident report, but the employee's supervisor is the party responsible for conducting the initial investigation. That supervisor knows more about the employee, the department, and the operation. Also, by definition the employee's supervisor is responsible for the safety of the employee and for department safety.

Accident Investigation Procedures

The Supervisor and the Accident Report

Accident report forms must be filed by department supervisors. They do this by asking the injured employee questions as well as talking to witnesses. Hold the supervisor accountable for completing the entire form. Every line should be completed, and all pertinent facts should be included. When the supervisor sits across the table from one of the employees and asks questions about the accident, it tells the employee "I am responsible for you, I care about how and why you get hurt, and I am answerable to top management to prevent this from happening again."

It is important to the injured employee that she or he be heard. Many times I find that supervisors paraphrase an employee's statement. Instead, an employee's statement should be in quotation marks and it *must be in the employee's words.*

This meeting gets the injured employee and his or her supervisor involved in the *prevention of recurrence.* The injured employee, more than anyone else, knows what caused the injury and can give you the information you need to prevent its recurrence. If the employee believes you will include him or her in the safety process, he or she will be much more forthcoming with information.

The supervisor needs to return the accident report to you on the same day the injury occurred. (If the injured employee cannot come back to work on the day of injury, then the supervisor needs to write the report based upon what facts are fresh in his mind and in the memories of witnesses.)

Investigate the Accident Yourself

When you receive a completed accident report, read the report carefully. Then, independent of the investigation that was held by the supervisor, reinvestigate the accident so that you can verify the facts.

Full-time Safety Directors investigate an accident themselves. Although you are not a safety professional or a full-time Safety Director, it is important that you visit the scene of the accident and get the basic information. The first line supervisor does an in-depth accident report, but you need to know what happened and be familiar with the equipment or process involved in the accident so you will be able to conduct the follow-up meeting.

The Accident Follow-Up Meeting

Call an accident investigation follow-up meeting that will involve the injured employee, the first line supervisor and you. It should be held within a minimum of twenty-four hours after the accident occurs and you receive the accident report. Tell the injured employee and the supervisor how concerned the company is about the accident, and explain that you are all here to resolve the problem that caused the accident so it will not happen again. Put the employee at ease and explain that you are fact finding and not fault finding.

Read aloud the supervisor's accident report and ask the employee for changes or revisions. Also ask the supervisor if there are any changes or revisions. If they both agree on the report, then ask them both if this was a preventable accident.

If they agree that the accident was preventable, ask what "we"—meaning the first line supervisor, injured employee, you and the company—must do to prevent this injury from occurring again. This is the real meat of the whole accident investigation: preventing recurrence. Injured employees need to know that they are part of the solution to preventing future injuries. If the employee was injured while committing an unsafe act, take this opportunity to go over the safe performance of the job. If the accident's cause was an unsafe condition, the three of you need to determine what action must be taken to correct this condition. An accident report without a follow-up sends a message, "The supervisor filed the report because he had to. Top management probably didn't even see the report, and nobody really cares." Send the message to the injured employee and his supervisor that injuries are a "big deal" and the company wants them stopped so that employees can work in a safe environment.

Real World: The Accident Investigation

The accident investigation means time, work, and effort on the part of supervisors. Often they don't see the real need for it. You will run into supervisors who will only partially complete the form, do a cursory accident investigation, and worst of all make meaningless, unreasonable or unworkable suggestions for prevention of recurrence.

When you receive an accident report that is not properly filled out, you can give it back directly to the supervisor and state the reason why, or give it to the first line supervisor's boss and state the reason why. Also inform your own boss of the reason for returning it.

Supervisors may try to intimidate an employee during the accident investigation report interview. They may say, "It's not that important. We need to get back to work." But they are really saying, "Don't rock the boat with me on this safety stuff or you will pay for it in the future."

When I run into cases in which I believe a first line supervisor is intimidating an injured employee, I then interview the employee without the supervisor present.

I tell the worker that it's my job to see to it that employees go home with the same number of arms and legs they came to work with, and that I need his or her help in doing this. I have found that many times an injured person will open up and state the facts as they actually happened.

Then I meet alone with the department supervisor and confront him or her with these facts. The first line supervisor must be made to know that I will not accept anything less than the facts and an honest effort in preventing injuries to employees under his or her supervision, which includes properly completed accident investigations.

Chapter VIII

"Let's take a look."

The Safety Inspection

Why do safety inspections? And who should be a part of the process?

Safety inspections allow you to determine the effectiveness of your safety program. An inspection tells you whether or not your program is growing or stalling, and in what departments. It tells your employees and supervisors that you are interested and are willing to get out of your comfort zone enough to go out and see if your policies are indeed being enforced. There is a correlation between departments that do well on safety inspections and good safety records.

Announced and Unannounced Inspections

Should you announce safety inspections? I have two schools of thought on the subject. My preference is the announced schedule inspection. I want supervisors and employees to know that they are going to be looked at and by whom. The reason is simple—I want to catch them doing something right!

It is easy enough to show up in a department and discover all types of discrepancies. But then, what is the message you have sent your department supervisor and the employees? Simply that you have the authority and power to make them look foolish, unprepared and incompetent.

In contrast, the scheduled inspection allows supervisors the time to go through the department and correct those things that they know they should have done, but hadn't got around to doing yet. It allows them the opportunity to show the department to you the way that they know you want it to be and to show you that they do care. The announced, scheduled inspection will also give you a strong message about the supervisor and about department safety.

It is possible that despite the notice that you were coming, the department will not be ready to be inspected. For example, personal protective equipment won't be worn, the housekeeping will be poor, walking and working surfaces will be unsafe, or employees won't be following basic safety procedures. The supervisor is telling you and the company, in front of the employees, that he or she doesn't think the inspection is important and doesn't care what you find. These "unprepared" supervisors want to see just what the consequences are, if any. Their unpreparedness is a message sent to you and it is one you cannot ignore. You will need to take action to correct the problems in the department and with the errant supervisor.

Using a Two-Inspection System

I suggest two types of inspections: informal and formal.

Informal Inspections

The department supervisor and one or two of the employees, on a monthly rotating basis, conduct a monthly informal safety inspection. In Appendix N is an inspection checklist for you to give them. They do a safety walk-through to discover what they can find. Their inspection checklist will help them with this process. Require them to turn in the inspection form to you on a monthly basis for your review. Be sure you actually do review it. It is always a good idea to visit a department or two every month to discuss with them the results of the inspection. Remember, "That which is measured, is done."

Formal Inspections

When you and members of top management conduct a formal monthly inspection of departments, it permits top management to mingle with the employees and to sell the safety message as well. In my work I have found it useful to group three or four departments into one area or zone that is formally inspected at least quarterly. That way all departments are visited at least annually by members of top management. The formal inspection sends a powerful message to the supervisor and employees: management cares enough to leave their offices and come out into the "real world" to see what is happening. Encourage the company's top management people to attend by advocating that this inspection is not only for safety, but also for quality and production.

The inspection for should be designed to express the idea of accountability to the department supervisor. The form should give a completion date that these discrepancies must be corrected by. It's up to the department supervisor to make sure that discrepancies are corrected. The deadline is very important. The supervisor returns to you the inspection report you sent him or her, with remarks stating that all discrepancies have been taken care of.

Then you must ensure that all the specified items really have been corrected. Be sure to follow through on this—the supervisor's acceptance of the inspection report procedure and that person's response to highlighted problems are important to building your credibility.

Real World: The Safety Inspection

Safety inspections mean more than a walk by you and other members of management through a department. Line employees closely watch to see whether you inspect in depth or go through a superficial exercise.

Stop to ask line employees questions about how safe they feel the department is, and what safety suggestions they have. A technique I have used is to ask at least one employee in each department, "What are we missing? What can hurt you that we aren't seeing?" A Safety Director's questions give the line employee a chance to open up some hidden issues that may not have surfaced before.

Asking questions of line employees in front of their department supervisor, your boss, or other top management demonstrates your rapport with the people you are charged with protecting, and their trust in you.

Although first line supervisors are responsible for ensuring that all discrepancies discovered during the inspection are corrected, they may not have to do the corrective work themselves. What they may be saying is, "I got the report and did the paperwork part, but if you want anything else done you'll have to go see the maintenance people yourself."

Don't let department supervisors put you into the role of the town crier. It is *their job* to communicate with the maintenance department to correct safety violations (Appendix M). When department supervisors are actively involved in the correction of the problem, they take ownership for the safety problems. That means they start to figure out ways to make their departments safer. However, it is a good idea to keep a "tickler" file to remind you to check on work orders for maintenance to monitor the situation from your end.

Chapter IX

"Tuesday isn't good for me."

The Safety Committee

Your safety committee is a vehicle you use to let those around you know what you are doing and to get input from them. I have seen almost every type of safety committee imaginable and find that in companies with over 500 employees, two safety committees are usually the best. For companies with less than 500 employees, one committee should suffice.

The safety committee is where your team is built.

Establishing a Safety Committee

Purpose of the Safety Committee

The purpose of this committee should be as follows:

1. To review new safety policies
2. To review safety training needs
3. To inform the committee of new federal and state standards

4. To review accident/incident data
5. To identify safety-related problems and their needed corrections
6. To review unsafe acts and unsafe conditions
7. To review accident investigation reports
8. To review anything that you, the Safety Director, feel needs to be reviewed, discussed or debated. This is your committee.

Members of the Safety Committee

1. *The operations director.* The operations director is the person who has the authority to get things done on the plant floor.

2. *The maintenance director.* In safety committees, a large amount of time is spent discussing what is broken, what needs to be changed and what needs to be installed or repaired. The maintenance director is needed at your committee meetings.

3. *The personnel director.* This individual deals with the employees and knows them. He or she can be your sounding board about the success or failure of what you attempt to accomplish.

4. *A member of top management.* Someone from behind one of the "big desks" needs to be there to see what you are doing, to give support to your efforts, and to ensure that the other members attend.

5. *Your insurance agent.* Yes, your insurance agent. You need to have him there to see the effort the company is making in reducing losses and improving its safety record. He is your voice with the insurance carrier.

6. *The company controller.* The company controller needs to hear and see where the dollars are going and the efforts being made to control them.

7. *The insurance company loss control representative* (Depending on premium). You need him there. It will depend on the amount of premium you pay as to how often he can attend, check with your insurance agent. This can be your best source of information and advise. He is probably going to be more knowledgeable than you and certainly more knowledgeable than the members of your committee.

8. *The insurance company claims representative* (Depending on premium). He handles your claims, which means he also handles where your money goes! This person needs to know the hard work and effort you are putting in to reduce losses. Again, premium paid will dictate if the claims representative can attend your safety committee meeting. Check with your insurance agent.

9. *One or two line employees.* Let the rank and file see what it is that you are doing. It gives them the message that this is for real and that they have input. These employees, who are recommended or selected by you, can do much to sell your program on the plant floor.

10. *You, the Safety Director.* You are the committee chair.

Running a Safety Committee Meeting

Don't allow the meeting to run more than an hour, or you will lose the attention of your members. You can accomplish your work in an hour if you are organized.

You run the meeting. Go down the agenda step by step (Sample agenda—Appendix T). Inform the members of what you are doing and what is going on. When an item on the agenda requires the input of a member, ask that person for his or her comments and suggestions. Give the impression you are in charge and that you will be making the decisions with their advice. *As you spend more time in your role, you will feel more comfortable in your position and will not need to be so rigid, but in the beginning it is imperative that you are the chair of the committee.*

I have found that many Safety Directors, at least initially, are tentative in their safety committee leadership. Usually they lose control of the group and are reduced to the role of scribe. Don't make that mistake! Come prepared with facts and figures, and remember—money speaks loudly, so talk money!

Take notes on the comments of the committee members. Don't become bogged down recording every word spoken by every member, but record the comments, opinions or suggestions that require future action due by the next safety committee meeting.

Before you end the meeting, state that "we will go around the table for opinions, suggestions, and comments." End with the most senior staff member.

Close by announcing when the next meeting is scheduled and who is responsible for doing what by the next meeting. Always re-state in the close who is responsible for doing what. Have a scheduled day and time every month for your safety committee meeting.

Safety Committee Follow-Up

I don't recommend typing up follow-up minutes of the meeting. It is extremely time-consuming and, quite frankly, most people won't read them. As your program develops you may then determine if minutes of the meeting will benefit your efforts. I do recommend sending follow-up reminders to committee members who have a task to complete before the next meeting. The follow-up reminder is crucial to keeping your committee on track. The method I have used is that two weeks prior to the committee meeting, I send a memo reminding the member of what he or she needs to do and inquiring as to what the status is on completion of that task. For example: "Joe, during our last safety committee meeting you stated you would check on the emergency exits in warehouse #3. I am in the process of writing the agenda for next month's meeting. Would you please give me the status of your exit inspection? Thanks, Bob."

The memo lets the committee members know you haven't forgotten what they are supposed to do and yet gives them a reminder to get the task completed prior to the next meeting or explain why it can't be completed. It is important to hold your committee members accountable for promises made in the safety committee meeting.

Union or Plant Safety Committee

A second kind of safety committee—the union or plant safety committee—is useful if your company has more than 500 employees.

This committee is made up of members from the plant floor who bring to your attention safety issues that need resolving and who discuss new ideas. This committee should consist of no more than eight members, and its function is strictly to listen to the issues of the rank and file to assure them that top management is listening.

Members of the Union or Plant Safety Committee

1. *Maintenance director*
2. *One or two supervisors*
3. *No more than five rank and file members* representing the different departments in your company.
4. *You, the Safety Director*. As in the safety committee, you are the chair, but unlike the safety committee you are the representative of top management. This is a committee to keep you in touch with what is going on in the plant and to address issues as they arise.

Especially in the initial stages, you pick the supervisor(s) who attend this committee. You want a supervisor who talks straight, plays fair and is respected by the plant employees. It doesn't have to be a popularity contest, but a well-liked supervisor can give great weight and strength to your efforts. You can then rotate as often as you feel necessary.

Running the Union or Plant Safety Committee

This safety committee should not last more than one hour; again, it is not an open forum or complaint session.

This committee is a listening forum. The representatives ask their questions or state their needs while you record them. Don't debate issues and/or attempt to resolve problems at this meeting. Stick to recording their concerns, *unless it is a case of imminent danger*, and if so, a supervisor should have already made you aware of the dangerous situation. (If the supervisor hasn't made you aware, correct the problem immediately and meet with the supervisor to determine why you weren't notified.)

The following month, open the meeting with answers to their questions. This gives you a month to do your homework.

If you really do get answers for your committee, you will build trust among the rank and file. It is important to them and to you that they know your word is good.

Real World: "Oh great, another meeting!"

Many people in your company will find your new role amusing. Attendees at your safety committee meetings will probably be there because they were ordered to. They won't know or won't care what it is you do; yours is just one more meeting in their already busy day. It will be even more difficult for you if you are a 5-foot female and if the majority of the committee members are male and have a higher rank in the company hierarchy.

You will probably be nervous and somewhat intimidated by their presence. So what do you do? Give yourself the authority you need!

- Get organized. Have enough on your agenda that it is clear you have "done your homework" and know what the problems are. Demonstrate that you have a good idea how to correct them (with the committee's input, of course).

- Let them voice their input as long as it's to the point and germane to the issues at hand.

- Get on mailing lists of your local safety council and other safety organizations and read their materials. You will give the impression of having the latest word in safety-related issues.

- Develop allies. Within the committee, look for advocates who believe in what you are doing. Talk to them before the meeting and enlist their support.

- Be assertive. You are the chair of the committee, you run the meeting! It's your show. Assertiveness is important because it makes the other members understand that they can participate, but that the Safety Director runs this committee. (Reread the passages in the introduction on being assertive if necessary.)

Chapter X

"You just got drafted!"

Putting Your Team Together

The first thing you need to remember is that you can't do it alone, nor should you be expected to. Successful safety programs are a team effort, with you performing the role of manager. You may not even be aware yet of the skills of your team members, skills that can help you achieve your goals. As with so much of your job as Safety Director, it's not always important that you have all the answers, but it is important that you know where to find them!

Your Insurance Agent

Top management is usually motivated to put a safety program into place because of the high cost of insurance. They are going to want to know if they are really going to see a dividend for their time, effort and money.

One of the key players on your team is your company's insurance agent. This individual is your bridge between the company and your insurance carrier. The agent can provide you with the technical information that you need, or can put the wheels in motion so that you get the information. Your company controller probably has a working relationship with your agent already, but you need to establish one of your own. It is you who will be selling the program not only to top management, but also to the first line supervisors and the line employees.

Information From Your Agent

You need to ask questions of your agent so you can get a general understanding of the insurance coverage you have and what you pay for each type of coverage. Examples of questions to ask and information you need are:

- What type of insurance plan are we on?
- What is our workers' compensation experience modification factor?
- What are our deductibles?

There are many more, but these give you a starting point. Sit down with the agent and explain that you need his or her help and ask that the agent explain to you the basics so you can be prepared.

The Agent Is Responsible to Your Company

You should also have expectations of your insurance agent in order to help you achieve your goals. The following are the responsibilities of your insurance agent to you and your company:

1. Keep your present insurance carrier updated on the operations and changes, if any, of your business.

2. Survey the insurance company marketplace to get the best cost and coverage possible for you.

3. Negotiate with insurance companies on your company's behalf to get you the best cost and coverage possible.

4. Assist your company in deciding what insurance company to use.

5. Assist you in the maintenance and growth of your Safety Program. This includes attendance at Safety Committee meetings and claim review meetings and participation in walk-through surveys of your company so as to be knowledgeable about company operations.

6. Review claims activity to effectively monitor the prompt and efficient handling of claims by your insurance carrier.

Your Insurance Company's Loss Control Representative

Your insurance loss control representative should be an immediate source of technical assistance. Some of the duties and responsibilities of the loss control representative are:

1. Help your company develop and implement an effective Safety Program that will reduce accidental loss.

2. Review your Safety Program to outline its strengths and weaknesses; advise you in ways to improve on strengths and to overcome weaknesses affecting assets and the health and safety of your employees and the public.

3. Use his or her knowledge and experience of accident causation and control to prescribe an integrated accident and loss control system.

4. Advise you regarding compliance with safety laws, codes, regulations and standards that apply to your operations.

Changing Loss Control Representatives

If after meeting with your representative you discover that this person is unable or unwilling to fulfill these four duties and responsibilities, you need to change loss control representatives. Call your agent and simply state that your expectations are not being met and you want a new representative to service your account. Remember, you are paying for this service and you should be receiving what you are paying for!

Your Insurance Company Claims Representative

The insurance company claims representative is your contact with physicians, chiropractors, physical therapists and other medical care providers who are treating the injured employee. You need to know what claims representatives do, and you need to define your expectations of them.

The claims representative should provide the following services for you:

1. Maintain cost-effective claim service.
2. Handle both medical-only claims and lost time claims. (Some insurance companies will have one individual for each type of case).
3. Investigate claims to determine compensability coverage.
4. Evaluate medical information.
5. Set realistic monetary reserves.
6. Contribute to vocational and medical rehabilitation through instructions to QRC (Qualified Rehabilitation Consultants) if your state has QRCs.
7. Give assistance to legal counsel.
8. Settle claims.
9. Attend conferences, settlements and hearings on your behalf.
10. Attend medical settlements.
11. Contact employer, employee and health care providers on claims.
12. Visit employer for on-site information.
13. Attend safety meetings and claims review meetings.
14. Review medical costs of health care providers.

The Real World: Putting Your Team Together

The success of your program depends in large part upon putting a good support team together. But there are a lot of blocks to successful coordination of insurance company input into your committee.

Start off with an "I don't know, I need your help" posture. After you get help and begin to learn the system, you can then raise your expectations and request (or demand) more. I know this sounds cynical, but it's simply the truth. The more you learn, the more you will discover what the members of your team know and are doing for you, or what the members of the team don't know and aren't doing for you. You are accountable to someone in your company for producing results; your team must do the same.

Meet the Agent

You need a good working relationship with your insurance agent. Don't be afraid to phone her or him to set up a meeting without the controller's knowledge, if the controller appears too entrenched in the organizational glue. I am not suggesting you be devious or duplicitous, I am saying one way or another you need to meet with the agent so you can do many of the things I have suggested in this chapter.

Because they themselves deal with the agents, controllers may not want you to meet with agents to discuss insurance issues independently. Your insurance agent may not know you. And your position, past or present, may be buried under so many layers of bureaucracy that you may not know much about the agent.

Most good insurance agents can sense internal discord in a company. The agent may call the controller to say that you have called and are wanting to know what's going on. Don't feel that you are being "back doored"—the agent wants to keep the account first and reduce injuries second. Companies with good safety records are easier to market to insurance carriers than those who don't have good safety records. That's business.

Agents come in many shapes, colors, forms and expertise. The one fear all agents have is *losing the account!* Therefore, the agent will want to please you and find out just what it is you need.

Two agenda items in the meeting with your agent should be the agent's explanation of the basics of your insurance plan, and your explanation of your expectations of the agent. The controller may want to get involved because he or she knows the agent and wants to know what you are up to. A technique I have suggested is to tell the agent and controller that the controller does the "risk financing," which includes the insurance. The Safety Director does the "risk elimination," which requires the participation of the insurance team.

Getting the Most from Your Loss Control Representative and Your Claims Representative

You are not expected to be a safety professional, but your insurance company loss control representative is. Like the insurance agent, this person doesn't want to lose the account. The insurance agent faces an immediate financial pinch if the account is lost; the loss control representative faces a negative phone call from you to your agent complaining of a failure by the loss control representative to provide the service you want.

Loss control representatives tell you they are there to help you reduce your losses. That's true to a point. They are there primarily to obtain underwriting information so the underwriters can determine premium and decide whether they want to continue to retain your risk. Loss control representatives will spend as much time with you as your premium dollar will allow. A company with a million-dollar premium will get much more service than one with a ten-thousand-dollar premium. You need to call your loss control representative to announce that you are starting a safety program and want to meet with her or him. Ask for professional help and find out just how much time the representative really can spend with you.

Your claims representative is someone you need to get along with, particularly if you are new to the claim management area. Most are overworked with a caseload like a New York social worker. They can make your job so much easier simply because they have the knowledge and experience in the field. Ask for their advice and counsel, but remember that they work for the insurance company and their primary loyalty is to their employer.

Chapter XI

"It costs how much?"

The Claims Review

The claims review committee is the place where you determine the status of your injured employees, their medical progress, and the cost to the company. It tells where your company's money is being spent by the insurance company and why.

Let's back up for a minute. Assume yours is an average company. How does the company handle its losses?

When an employee is hurt, the insurance company pays the bills and the matter, after a minor investigation, is usually forgotten. Employees who suffer a lost time injury are off work until the doctor releases them to work. The insurance company's claims adjuster is usually the only person to speak with them, and that is to ensure that they are getting his checks. Your company maintains no contact with the injured person, doesn't really know where the employee is in terms of recovery, and usually has no idea how much the injury is costing.

This attitude that says "the insurance company is handling it" is all too common. But it puts the insurance company in control of your injured employees, your dollars and, for all intents and purposes, you, the Safety Director. The insurance company has the facts about the situation and you don't!

Purpose of the Claims Review Committee

The function of the claims review committee is to determine these facts: Where do we stand? What outstanding claims do we have? What have we paid, and what is the insurance company reserving and why? When was the last time you spoke with the doctor? The employee? What is the prognosis? When will reserves go down? What are our options?

Members of the Claims Review Committee

Who should attend a claims review committee meeting?

1. *Your insurance agent*—In a safety program there can be no greater source of help to you than your agent. Keep in mind that the agent works for you and is responsible primarily to you and your company. The agent is responsible for getting you the best insurance carrier with the best price and the best service. If, however, you show little or no interest, he or she can only assume that things are going smoothly. The agent must keep the bridge open between you and the insurance carrier, but the insurance agent is just that—your agent.

   By inviting the agent to the claims review, you are sending the message that you expect answers, not only from the insurance carrier, but also from the agent personally. Your agent can be a source of great support to you in the claims review by interpreting why the insurance company is doing what they do.

2. *Insurance company claims representative*—The most important person on your team to assist you in keeping track of your insurance claim dollars is your insurance carrier claim representative. This person can tell you the employee's medical

status and the amount the claim has cost your company to date, plus give you advice on how to handle a particular case.

In the event of a contested workers' compensation case, it is the claims representative who, along with your insurance carrier's attorney, will represent your company at state workers' compensation hearings. A claim representative can be a source of great assistance to you and can make your job run much smoother.

In almost all cases, claims representatives work without much input from their insured. Most of their client companies do not have a monthly claims review to determine the status on each and every one of their cases. It may come as a surprise to your claims representative that you expect him or her to show up monthly or quarterly to provide you with assistance and information. The frequency of visits and quantity of service will depend upon the amount of premium dollar you are paying. A million-dollar yearly premium will receive much more service than a ten-thousand-dollar yearly premium. Talk to your insurance agent to determine a realistic plan for visits from your claim representative.

Claims representatives are probably going to be a little hesitant and defensive coming into your claims review, at least at first. Their mindset says, "I have been handling these claims with no problem and now I have to answer to this customer who probably has no knowledge of what I do. I get audited by my department manager and my home office, and now I'm audited again by the account."

It is important for you to let them know you are only trying to determine how the claims are being handled and why. Make sure in your own mind that the insurance company works for you as a vendor supplying a service. You need the claims representative on your team and you want a good working relationship with that person.

If the claims representative wants to call the shots without your input, you need to tell your insurance agent you are not satisfied with this arrangement. The insurance carrier does have the right to determine how to handle claims in their best interest and yours. But as the customer, you have a right to give input. Your insurance agent is the person who addresses your concerns with the insurance carrier.

With all of that said, I wish to stress the importance of trying to avoid an adversarial role. Your insurance claims representative is a very strong element in your safety program.

3. *The operations director*—Have your operations director in your claims review. When you are discussing employee injuries, it is his or her employees you are discussing. When you talk about the cost of an accident/incident, it's important for the operations director to know what the expense was, not only in terms of countable dollars, but uninsured costs such as downtime for accident investigation, etc.

4. *The controller*—Money talks, so talk money. This person needs to know where your dollars are being spent by the insurance company and to evaluate whether the claims dollars are being handled properly.

5. *A supervisor, on a rotating basis with other supervisors*—Each first line supervisor needs to know the consequence of an accident in his or her department. I have found that supervisors have no idea of the cost of a laceration or back injury and they are shocked to see where the money went. They must be part of your claims review as a learning experience—have at least one supervisor there.

5. *Member of top management*—They need to see you controlling the company's dollars and losses. This will be an educational experience for them as well. The presence of top management also provides you with immediate authority to make decisions as to what to do with your injured employees or with claims that need a decision.

Putting It All Together

Through the claims review committee, you and other members of management find out the status of injured employees and what the cost to the company is for these injured employees.

As the chair of this committee, you will bring up each injured employee's name and the claims representative will give you the status. The claim representative may assign QRCs (Qualified Rehabilitation Consultants) to your lost time cases depending on your state law. The insurance company pays for this consultant service, which means you are paying for it. You may want to ask what the QRC report indicates about bringing an injured employee back to work on modified duty, and what the general attitude of that employee is about coming back to work. Again, the whole purpose of this committee is to keep you and other members of management informed of the claim status on your injured employees.

The Navigate Program; Return-to-Work

When I took over as Safety Director, the first thing I discovered is that we had fourteen employees on workers' compensation, and the only people who even knew where they were was the claims representative who mailed their checks to them. That's not uncommon. An insurance company is happy to send workers' compensation employees a weekly check and not notify you—after all, it's your money.

Meet with your claims representative and agent to find out what your workers' compensation cases are costing you. Look at the dollar amounts that are lost if you don't bring your workers' compensation employees back, and the costs if you do bring them back.

The 100% Rule vs. Modified Duty

Many companies have the 100% rule. If employees cannot function at 100% then they stay home until the doctor, usually their personal physician, releases them back to work. This system is clean, neat, causes few personnel problems, and is very, very expensive! Sit down with your agent and claims representative and go through the cases that you have to see what it is costing your company to indulge in this luxury.

Find out where those dollars are going and compare that amount to what it would cost if you brought the employees back. Have your agent and your claims representative work up cases for your review so you can be armed with the knowledge you need to present to top management.

In the area of workers' compensation, nothing has caused more turbulence than the issue of whether or not to bring an employee back to work to a *modified duty job*. Notice that the term is "modified duty," not "light duty," or "restricted duty." Those terms give the impression of a free ride. Use the term "modified duty"!

When you approach management with the issue of bringing employees back to work on modified duty, you need to be prepared by learning what it costs the company to leave them home.

Learn to Navigate

The Navigate program is your approach of dealing with your injured employees. "Navigate" means that since the injury, you are not going to allow your injured employees to "drift" along in the workers' compensation system. Too many companies suffer high workers' compensation premiums simply because they don't get involved and they don't make the effort to "navigate" the return to work of their employees. They don't follow-up with the employee, the doctor, or the claims representative. Your involvement in navigating your injured employee is very important to early return to work.

If putting a still-injured employee to work were easy, everyone would be doing it! Convincing top management, first line supervisors and the injured employees to bring them back to work sounds simple conceptually, but functionally it's difficult. You will hear, "Where do we put them?"; "It will cause more problems,"; "It will allow a free ride for the injured employee," etc. In many cases these statements are true. Nevertheless, you need to make a decision—and then make the recommendation to top management—one that your company cannot afford the luxury of paying higher dollars on insurance premiums just to have your employees sit at home.

Modified Duty: Putting the Injured to Work

Never allow an employee to bring you a generic doctor's slip saying he or she should be placed on "light duty." What is light duty—lifting no more than ten pounds? No repetitive work? No sitting? No standing? This generic light duty excuse is totally useless for your purpose. To define the modified duty that each injured employee can do, you must begin with a doctor's form that states what the employee can and cannot do! It gives you some ideas where and when the employee can perform work. (See the doctor form in the Appendix R.)

The Medical Consultant

One of the very first things the Safety Director will need to do is to find a company medical consultant who specializes in occupational injuries, or is willing to work with you on occupational injury cases. Notice I didn't say company "doctor"—that term sets off alarms—but instead used the term "company medical consultant."

Your selection of this physician is very important. This doctor must understand your desire to return your employees to work as quickly as medically possible. You want an occupational medicine doctor, one who knows that not all work-related injuries automatically mean lost time. You need to sit down with the physician and explain your return-to-work policy. A company medical consultant that sends your employees back to work "no matter what" does you more harm than a physician who gives ten days off for a minor scratch.

The medical consultant needs to have the trust of your employees and your confidence in his or her objectivity. Some questions to consider when choosing a medical consultant: Does this doctor make referrals? Does the doctor attempt to build a rapport with the injured employee? Can an employee trust this doctor?

Included in the Appendix is a medical capabilities form and a cover letter for a doctor. Your company medical consultant will use the first form, and the cover letter will be used to inform the employees' own physician about your policies.

Getting physicians to fill out this form can be a real challenge. This is another situation in which your claims representative needs to be in contact with the doctor. Let the doctor know that part of the medical treatment for the employee—which the insurance company is paying for—is completing the form. You and the claims representative need to be assertive in this activity.

Another source for you to use in establishing return-to-work as the norm is to use a return-to-work firm that charges you on a case-by-case basis to follow up on the employee and that employee's visits with the doctor. I have used return-to-work firms and other medical management companies and found them very useful in getting reluctant employees back to work.

Visit with the Doctor

Whether it be you or another company representative, it is important that someone who knows the injured employee's job sit down and explain the job to the doctor. Why? Well, the classic example says:

> **Employee:** "Doctor, my back is killing me. All day long I must lift 100 pound bags of rock salt and I can't keep doing it with an aching back."

> **Doctor:** "Isn't there another job you can do?"

> **Employee:** "No, my job is lifting 100-pound rock salt bags."

The employee is not lying to the doctor; he is required to lift 100-pound rock salt bags all day. However, the employee has failed to mention that he is a forklift operator and lifts the pallets of rock salt with the forklift.

Sometimes the employee is not being deceptive, but is unaware that your company will find him or her another job within the medical restrictions. Someone, whether it be you or another company representative, needs to tell the treating doctor that you can provide modified duty that will fall within the injured employee's medical restrictions.

Modified Duty and the Supervisor

Finding work for the modified duty employee can be difficult, especially if the first line supervisor is fighting the return-to-work concept. The perception of the first line supervisor and department employees may be that the modified duty employee will not be doing a fair share of the assigned work.

Modified duty is based upon what the employee can do rather than what the employee can't do. For example, Can the employee sit at a bench and sort parts or do bench assembly work? A good method of finding where you can assign an employee is to review your job descriptions as to physical activity required. What are the weight restrictions for the jobs in your company? Knowing this information will allow you some freedom in attempting to assign your injured employee to a modified duty job.

You must educate your supervisors on the need for their involvement in this process. Talk money: show them the cost of leaving the employee at home as compared to bringing him or her to work. Explain to them why the company has no choice but to take this course of action. Explain that it will be a fair and equitable program: all supervisors will be required to have their share of modified duty people, and they will all be expected to support management's decision to assign these people to them.

Many return-to-work/medical management companies will conduct a training program for your supervisors so they will understand what is going on. Ask your insurance company or agent for names of return-to-work firms.

Real World: The Claims Review and Return-to-Work

In the claims review, you deal with egos outside of your company. The claims representative is a key player. You must make it clear to the claims representative, from the very beginning that they answer to you first and the committee as a whole second.

You are in charge. Call the claim representative and your insurance agent to set up an initial meeting. Tell the two of them how you are going to structure the claims review and how you are going to run it. You will bring up each claim and ask the claim representative what progress is being made on medical recovery according to the doctor's report, and whether the insurance company is still charging your company through reserves on this injured employee. Although every committee member is there to be briefed on the status of claims against the company, the claims representative answers to you at the meeting. The return of injured employees to work can mean conflict at the meeting. Ask the claims representative to tell you and the committee what it is going to cost to keep the employee off work, but you don't have to put yourself in the position of saying yes or no on the return-to-work issue. Many department heads will try to get you on record as the decision-maker so they can say "it wasn't my decision." But the decision will be made by you, your boss, and others in top management, and the deciding factor will be the economic cost to your company. Don't feel forced to make a decision at the meeting. Simply make notes, and say that a decision will be forthcoming.

Don't be maneuvered into the position of wearing the black hat on this one. It's your job to prevent injuries *with the cooperation of all company personnel*. Also, it's your job to determine what the economic cost to your company will be if you don't bring workers back. And it's the job of you, other department heads, and first line supervisors to see to it that these modified duty employees are mainstreamed back into your work force. You are only part of the solution; you can't do it all and you shouldn't allow others to paint you into a corner where you are expected to. It will bog you down in solving your own safety issues.

Chapter XII

"A friendly visit from OSHA."

You and OSHA

Nothing strikes fear into the heart of a Safety Director more than hearing the receptionist tell you the OSHA inspector is here to see you. Don't panic—it's not that bad!

Preparing for an OSHA Inspection

What Is OSHA?

OSHA, the Occupational Safety and Health Administration, is a federal agency within the Department of Labor. It is the outgrowth of the Williams-Steiger Occupational Safety and Health Act, passed by Congress in 1970. The purpose of the law is:

> "To assure safe and healthful working conditions for working men and women; by authorizing enforcement of the standards developed under the act, by assisting and encouraging the states in their

efforts to assume safe and healthful working conditions; by providing for research, information, education and training in the field of occupational safety and health; and for other purposes."

There can be two OSHAs, depending on the state where you live. States can have their own OSHA program and their own compliance officers if approved by federal OSHA. They are monitored by federal OSHA to ensure they are as effective as federal OSHA. In states that do not have their own state OSHA, the Occupational Safety and Health Act is administered by the federal government.

Learning What OSHA Standards Are

1. If you do not know if your state has its own OSHA inspection plan, simply call your local safety council or federal OSHA to ask.

2. Write down the phone numbers of the federal OSHA office and of your state offices, if applicable. Keep these phone numbers handy because you will be using their services when you are unsure of something or when you get questions you can't answer.

   You aren't expected to know all the regulations that apply to safety, but you will be expected to know where to go to get information. Questions often arise when you don't have the time or knowledge to do research. "What is the standard interpretation dealing with forklift operations? Just what does 'trained and authorized operators' really mean? Do employees have to attend a forklift school or will a training video suffice? What about employees who refuse to perform a job because *they* think it is unsafe—is there an OSHA standard referring to this?"

   It is more than all right to call OSHA and tell them you're confused, bewildered or overwhelmed—it's smart! They know the standards and they will interpret them for you.

2. You need to know the rules to follow. These rules are referred to as "standards." The type of business you are in depends on the type of standards and rulebook to follow:

a. General Industry: 1910
b. Construction: 1926
c. Agriculture: 1928
d. Maritime: 1915, 1917, 1918

Call your local OSHA office to inquire as to how you can get the book of standards applicable to you. Usually you can order the federal standards book from the federal government, or your state OSHA office can give you the standards for both state and federal.

The standards—rules and regulations—can be overwhelming when you first open your book. For this manual I will use the "1910" standards book for general industry as the model.

How do you know where to look in the *1910* for the applicable code, and is it relevant to your situation? Where can you get help in finding the right regulation without looking like you don't know what you are doing? You need to research answers for questions that will come up from time to time and have the answers ready so you give the impression that you are at least conversant with the standards.

The first thing you need to purchase is an OSHA reference manual. OSHA standards are black and white. You are either in compliance or you are not!

Your next source of OSHA information is your insurance company representative. A simple phone call requesting the standard and your representative's appraisal of whether you are in compliance is part of the loss control service that is part of your premium. This is a service many Safety Directors have but don't use. *Don't accept "I don't know" for an answer.*

If your loss control representative can't get you the needed information, a phone call to your insurance agent will start the wheels moving to get your questions answered. Again, you are paying for this service and you should be getting it.

The third source of OSHA research assistance is OSHA itself. You can call your local OSHA office and ask the compliance officer as many questions as you want. If you prefer you can remain anonymous, although that is really not necessary as they don't write down the names of people calling and put them on their inspection list. If you are uncomfortable with saying you want to remain anonymous, then use a fictitious name. OSHA can give you the specific answer, chapter and verse if you will let them work for you.

I have been researching OSHA codes for the twenty plus years OSHA has been in business. I don't know all the standards and don't have unrealistic expectations of myself about knowing them all. You shouldn't either! It's not important that you know them, it's only important that you know where to look or who to call to find out. That is the way safety professionals operate and it's the way you can perform your role as well.

Don't Resist the OSHA Inspector

You are not going to be notified that OSHA will be visiting your company. In most cases, your receptionist will just tell you that the OSHA inspector is waiting in the lobby.

Your best preparation for an OSHA visit is to handle your record keeping as if OSHA were coming today. Keep your OSHA 200 log up-to-date. Keep all OSHA logs for the past five years easily accessible and readable. Be sure to have your files ready for the compliance officer to review. Look organized!

When the OSHA compliance officer shows up, be there to meet the person. Be courteous and professional. The compliance officer is going to be calling the shots—let it happen. Make it clear that you know that he or she is in charge and that you are more than willing to comply with his or her request.

The compliance officer will be inspecting your business for violations of standards. If one is found, a citation will be written and mailed to you. OSHA will give your company an amount of time to correct the violation—an abatement.

Don't convince yourself OSHA is out to get you and should be considered the enemy. Contrary to popular belief, the compliance officer doesn't get a percentage of any fines brought against you. The majority of the officers I have met are people doing their job. Treat them with respect, as you would any professional.

You can refuse to admit an OSHA compliance officer and force her or him to return with a court order. Put yourself in the position of the compliance officer. There are many ways that a company under inspection can hurt itself—this is one of the best.

The compliance officer may be visiting your business because of a complaint from an employee. If this is the case, do not attempt to interrogate the compliance officer about who the complainant is or why the complaint was made. Don't discuss what a terrible, malcontent employee this person is. Be a professional; the complaining employee may have a legitimate complaint. Remember, you will either be found in compliance or you will not.

The compliance officer probably has been in the business far longer than you have. Listen to this person's advice and counsel and make notes about problem areas that are pointed out. *DO NOT*—repeat, *DO NOT*—make an adversarial encounter. *Make it blatantly obvious to the compliance officer that you want to learn, are eager for suggestions and are more than willing to comply with whatever findings are made.*

While OSHA Visits

Whether it is state or federal OSHA, the compliance officer who visits you will first show you some kind of Department of Labor identification. Then he or she will want to have an opening meeting with you to state why your business was selected and will ask that an employee of the company accompany you on the inspection. A line employee who is a member of the safety committee is a good choice, or a union representative may go along.

You will then accompany the officer on a tour of the business. You are not the tour guide and you are not in control. The compliance officer will set the pace and stop and ask questions of your employees. Don't answer for them and don't interrupt. Let the OSHA compliance officer do her or his job.

You will be questioned about your safety program and what you are doing to comply with OSHA standards. Don't state something if you are unsure or don't know. Say that you don't know, but will find out. Don't try to impress the compliance officer with your great knowledge of OSHA standards. You'll only be wasting your time. To use a cliché, this person "has forgotten more about OSHA standards than you will ever know!"

The OSHA inspector is the teacher and you are the student. Take a notebook along and make notes to show a sincere interest in the improvement in the safety of your business. Treat the compliance officer with respect and courtesy: he or she is a professional doing a job and deserves as much.

When your inspection is over, you will then return to your office for a closeout. This is the time when the compliance officer reviews the findings with you and discusses possible violations. The dollar amount of penalties will not be discussed at this meeting. The compliance officer needs to review his or her findings with a supervisor to determine which citations will result in monetary penalties. These results will be mailed to you or to the authority you tell OSHA to send them to. You should have them mailed directly to you for your review and subsequent corrective action.

This closeout meeting is important in conveying an impression to the compliance officer. You need to send the message that you are interested and sincere, and that you want to improve your safety program. Ask questions about the findings. Take more notes, and ask advice on how to correct a problem. Use this opportunity to gain knowledge and to convey a positive image of your company.

Six Types of Violations

There are five major types of violations, plus one that is not as devastating to your company.

1. *Willful Violations*—A "willful" violation may exist under the Act when OSHA determines that you knew there was a hazardous condition that could kill or cause serious injury and didn't correct it. A willful violation can cost you up to $70,000. (If death results from a willful violation, the violation is in the category of *Criminal Willful Violations* and can cost you more money, plus imprisonment.)

2. *Repeated Violation*—Repeated violations are cited if you have been inspected and a violation was discovered, and upon reinspection, the same violation or a similar one is found. For a repeated violation, you can be fined up to $70,000.

3. *Serious Violation*—A serious violation must have the potential of causing a serious injury or death, and it must be obvious that the company should have known of the hazard. A serious violation can cost your company up to $7,000.

4. *Other than Serious Violation*—This violation won't cause death or serious injury, but it is an infraction of the standard. Here is where the compliance officer and the officer's supervisor can reduce the maximum fine of $7,000 to as little as 20% of the maximum, depending on your safety record and your commitment to a safe workplace.

5. *Failure to Correct*—If a violation is cited in the failure to correct category, then $7,000 per calendar day beyond the abatement date is assessed until an OSHA inspection determines that you now comply with the standard.

6. *De Minimus*—De minimus violations are minor infractions of the standards. They will be included in the report, but there is no fine for them.

According to OSHA, repeated violations differ from willful violations because they are the result of an inadvertent, accidental or ordinarily negligent act. A repeated violation may actually be willful, but the element of willfulness may not be sufficiently provable.

101

The key to weathering an OSHA inspection is to look prepared, interested, sincere and organized. An OSHA compliance officer will usually give more leeway to a Safety Director who demonstrates those traits than to an individual who does not.

Real World: You and OSHA

OSHA will find something wrong in your company when they do a compliance inspection. They are there to find something wrong, that's what they are paid to do. They also are not stupid – they will know in a minute if you are trying to impress them with your skill and knowledge: compared to them you don't possess either. OSHA wants to see records documentation, something in writing. Remember, *if you didn't write it down, you didn't do it*. OSHA is looking for the obvious, the things you should have known to have done. Answer their questions as honestly as you know how—tell them the truth and show them what they ask to see. Don't volunteer any more than they ask for and don't take the compliance officer to areas where you are having problems. Use OSHA consultants for that. In dealing with OSHA, keep your mouth shut, ears open, look sincere and interested, and have your record keeping up-to-date.

Chapter XIII

"What do I do now?"

How to Survive an OSHA Inspection

In the previous chapter on OSHA you were given a general overview on the who and what of the Occupational Safety and Health Administration. In this chapter I want to deal in specifics—the specifics of How to Survive an OSHA Inspection. There are techniques that will assist you when you find a Compliance Officer waiting patiently in your lobby.

Tips for Dealing with the Compliance Officer

1. *Be prepared for an inspection today.* As easy as it is to say that, the reality is that being prepared all day, every day is indeed difficult. Veteran Safety Directors, however, know and understand that OSHA will not be sending you a letter in advance. Treat your mandated recordkeeping and training as if they did give you advanced notice. Have your 200 log up-to-date, as well as your Job Safety and Health Posters up where employees can see them. Understand that when OSHA does visit, and cites you for violations

you could have easily prepared for, you will soon learn to prepare in advance.

2. *Conduct periodic self-assessments.* Use OSHA consultative services or private Safety Engineering Consultants if you are new to your position. You need someone with experience to do an assessment of your company. You cannot be expected to know all there is your first week on the job. Your boss may think so, but it's just not going to happen. Get help from somewhere in order to discover the unsafe conditions and standard violations that exist.

3. *Develop a written company safety policy.* In this book is an example of a company safety policy (Appendix C) I wrote many years ago. I have seen it used in companies all over the country. Have the senior management official on-site sign it and post it in the front lobby. Let OSHA know when they walk in the door that your company is committed to employee safety.

4. *Train employees on safety.* Nothing you will do in your position is as important as employee safety training—nothing! OSHA is more flexible with employers who have demonstrated good faith intention through training. Again, nothing is as important as employee safety training!

5. *Do not refuse access to the Compliance Officer.* I often refer to this as 'Felony Stupid.' Recent statistics have shown that employers who refuse access are cited more frequently, more severely and with higher fines. If you are prepared, courteous, and cooperative you have a much better opportunity than being unprepared and belligerent.

6. *Develop a rapport with the Compliance Officer.* Think of a Compliance Officer as a State Trooper or IRS Auditor. What does it benefit you to become hostile and combative? Remember, in the CFR 1910 there are over sixty-five thousand standards you can be cited for. Do you really think that your company is in compliance with each one? A positive, cooperative attitude in the opening conference will assist you and your program in the inspection process.

7. *Ask for credentials.* Be on the look out for sales representatives posing as Compliance Officers. They have used this ploy in the past to gain access to businesses. State and Federal Compliance Officers carry proper identification and usually present it when they enter.

8. *Ask to see the complaint.* If the compliance was prompted by a complaint, ask to see it. The Compliance Officer will not tell you who lodged the complaint, but you are entitled to know what has been alleged.

9. *Invite other parties along on the inspection.* Plant managers and department supervisor can assist you in answering questions regarding specific technical questions. If you have a union, have the steward go along as well to represent the employees. If you are non-union invite an employee or two along to demonstrate your commitment to their involvement in the program.

10. *Do not try to steer the Compliance Officer during the walk-through.* Conduct the opening conference in your office or conference room. After that allow the Compliance Officer to do the walkthrough as they deem necessary. These people are not stupid. They will immediately know if you are attempting to navigate them—do not do it!

11. *Employee interviews are confidential.* The Compliance Officer has a legal right to interview employees in private without a representative of the company being present. Give the Compliance Officer credit for knowing how to distinguish valid complaints from those that are not. Nothing will alert a Compliance Officer faster than your protestations about a certain employee. The standards are very clear—you are either in compliance or you are not.

12. *Make your own record of the inspection.* If the Compliance Officer takes pictures, sound level measurements, or video taping of the inspection you should do the same thing. You may need that information later to contest citation. Do this without a contentious or antagonistic attitude. Let it be known that you are so interested that you to want to record these events.

13. *Make effective use of the closing conference.* Be sure to request a closeout meeting with the Compliance Officer after the walk-through if he does not suggest one. This is the time for you to hear him present potential violations and characterizations of violations. You may be able to negotiate a reduction in severity or abatement requirements at this level. You may also be able to satisfy the Compliance Officer's concerns on questionable items and avoid a citation altogether. At the very least you will learn the basis for a citation so that you can start preparing your defense. *Do not nod and agree that you are in violation. Simply say that his comments are noted.* There have cases in appeal when the Compliance Officer has testified that the Safety Director agreed with his assessment of violations. Again, simply state that his comments are noted.

14. *If you need more time to abate, ask for it.* The Compliance Officer usually has some discretion on the abatement period for hazards that do not pose an imminent danger. If you are cited for a condition that will be difficult to abate, discuss your needs and the potential difficulties with the Compliance Officer to justify a more appropriate abatement period.

15. *Do not forget to post the citation.* The Compliance Officer will usually follow-up with employees to make certain the citation was posted as requested by OSHA. Remember, additional fines can be imposed for failure to post.

16. *Notify OSHA that you have abated the violation.* Additional penalties can be imposed for failure to abate a violation. Make sure OSHA knows that you have corrected the condition that prompted the citation, even if you contest the citation itself.

17. *File your protest within the time allowed.* You must file your protest usually with 15 working days of receiving the citation, even if you intend to request an informal conference or settle the citation without further appeal.

Please Note: Be sure to learn your state plan time limits on filing protests. Do not forfeit your rights to contest the basis for the citation, characterization of the violation, amount of the fine, and abatement period.

The Appeal Process

The Decision to Appeal

Let's now address the decision to appeal. For most Safety Directors the decision to appeal an OSHA citation can be difficult since your expense of legal representation may often appear to outweigh the cost of the fine. In many cases your boss will ask your advice on the matter and you need to be able to state all the variables available to your company. Some of the factors you need to consider are:

1. The true cost of an OSHA citation can often be much greater when the long-term ramifications are considered.

   A. Your company's record of violations can often result in future citations being characterized as "repeat offenses" with higher fines. Generally speaking, a repeat offense is one in which occurs within three years of a prior offense involving "substantially similar conditions."

   B. An uncontested citation can sometimes be used as evidence of your company's negligence in the event an employee is injured.

   C. Since OSHA violation records are available to the general public through the Freedom Of Information Act, they can be used by competitors, unions, or the news media to create misleading impression that your company is unconcerned about its employees safety.

   D. Failure to contest violations can earn your company a reputation as an "easy mark" by over zealous-inspectors from a variety of compliance agencies, discontented employees, or trade unions.

2. In many cases, abatement of the citation and penalty can be negotiated after a notice of contest is filed.

   A. Often OSHA has limited resource to devote to the hearing process, and hearings take inspectors out of the field.

   B. OSHA has the burden of proving the validity of the citation and reasonableness of the proposed penalty and the abatement period.

   C. In many cases they can be persuaded to drop citation with inadequate legal factual, or scientific justifications.

   D. OSHA can also be persuaded to compromise when it is unable to demonstrate a practical abatement method exists for the hazard cited.

   E. Your company often has an opportunity to negotiate a settlement that does not require an admission that a violation occurred.

   F. Remember, negotiated settlements can help the company avoid being labeled as a repeat violator.

Keep your options open and inform your boss that he or she should at least consider the need to contest OSHA citations. Giving your boss information and options is part of your job. Let's now consider that your boss has decided to appeal the citation, now is the time to inform him or her of the Appeal Process.

Federal OSHA Appeal Process

A. *Appeals by Employees.* If an inspection was initiated due to any employee complaint, the employee or authorized representative may request an informal review of any decision by federal OSHA not to issue a citation.

   1. Employees may not contest citations, amendments to citations, penalties or lack of penalties. Your employees may

contest the time specified in the citation for abatement of a hazardous condition. They may also contest your company's Petition for Modification of Abatement (PMA) which request an extension of the abatement period. Your employees must contest the PMA within 10 working days of its posting or within 10 working days after an authorized employee representative has received a copy of the PMA.

2. Within 15 working days of your company's receipt of the citation, the employee may submit a written objection to OSHA. The OSHA Area Director forwards the objection to the Occupational Safety and Health Review Commission, which operates independently of OSHA. Your employees may request an informal conference with OSHA to discuss any issues raised by an inspection, citation, notice of proposed penalty or employers notice of intention to contest.

B. *Appeals by Your Company.* When issued a citation or notice of a proposed penalty, your company may request an informal meeting with OSHA to discuss the case. Your employees or their representative may be invited to attend the meeting. The OSHA area director is authorized to enter into settlement agreements that revise citations to avoid prolonged legal disputes.

C. *Petition for Modification of Abatement (PMA)* Upon receiving a citation your company must correct the cited hazard by the prescribed date unless your company contest the citation or abatement date. If factors beyond your company's reasonable control prevent the completion of corrections by that date, your company, which has made a good faith effort to comply, may file a PMA for an extended abatement date.

1. This written petition should specify all steps take to achieve compliance, the additional time needed to achieve complete compliance, the reasons the additional time is needed, and all temporary steps taken to safeguard employees against the cited hazard during the intervening period. It should also indicate that a copy of the PMA was posted in a conspicuous place at or near the each place where a violation occurred,

and that the employee's representative (if there is one) received a copy of the petition.

D. *Notice of Contest.* If your company decides to contest either the citation, the time set for abatement, or the proposed penalty, you have 15 working days from the time the citation and proposed penalty are received in which to notify OSHA in writing. An orally expressed disagreement will not suffice. This written notification is called a "Notice of Contest."

1. There is no specific format for the Notice of Contest; however, it must clearly identify your company's basis for contesting the citation, notice of proposed penalty, abatement period, or notification of failure to correct violations.

2. A copy of the Notice of Contest must be given to the employees' authorized representative. If any affected employees are not represented by a recognized bargaining agent, a copy of the notice must be posted in a prominent location in the workplace, or else served personally upon each unrepresented employee.

Appeal Review Procedure

A. If the written Notice of Contest has been filed within the required 15 working days, the OSHA Area Director forward the case to the Occupational Safety and Health Review Commission (OSHRC). The Commission is an independent agency not associated with OSHA or the Department of Labor. The Commission assigns the case an administrative law judge.

B. The judge may disallow the contest if it is found to be legally invalid, or a hearing my be scheduled for a public place near your workplace. Your company and your employees have the right to participate in the hearing; the OSHRC does not require that they be represented by attorneys.

C. Once the administrative law judge has ruled, any party to the case may request a further review by OSHRC. Any of the three OSHRC commissioners also may, at his or hew own motion, bring a case before the Commission for review. Commission rulings may be appealed to the appropriate US Court of Appeals.

As overwhelming as all of this sounds, it is the type of information that you need to know and to make available to your boss. Be sure to call either federal OSHA or your state OSHA to ensure that their policies and procedures have not changed since you have read this material. There is nothing wrong with contesting a citation if you honestly feel that there are extenuating circumstances that need to be addressed.

Chapter XIV

"Turn left at the light."

Following Your Safety Program

You're now ready to put your own program together. In the previous material, we've covered several important concepts. To assist you in putting these together, this chapter presents an outline of the Safety Program with commentary. A condensed version of this outline can be found at the end of the chapter.

Remember, this program serves as merely a guideline. It is not written in stone! The safety program is *your* program. As the Safety Director, you can make as many changes as you want. You have to map out a program that takes into consideration your company's unique environment. And always remember, what you do is important. Good luck!

I. Management

A. Safety Policy Statement

1. The company has a general policy on its commitment to accident prevention.

A policy statement on safety is included in Appendix C. Feel free to use it and to make any changes you feel necessary.

2. The policy is signed by the chief executive officer.

You need to go as high in executive management as you can get for the endorsement of your policy. Your company's employees need to know that this policy represents the thinking and commitment of the very top level of your company's hierarchy.

If your CEO or president is not on-site or involved in your day-to-day operation, you will then need to get the signature of the on-site authority who is actually in the building and is seen and heard by your employees.

3. Policy statement is distributed as follows:

a. Each department manager

Your department managers need to know that this safety policy is indeed "company policy" and that it represents the posture of the company. They need to know that they will be held accountable for their adherence to the policy and that there will be consequences if they do not adhere to it.

b. Each first line supervisor

First line supervisors must have a copy in their hands. They need to know that they are responsible for enforcing this policy and what the consequences are for not doing so. *Remember, the first line supervisor is your real target for this policy.*

c. Enlarged and posted in main lobby

Use the posted policy as a device to influence others' perception of your commitment. Tell the world of your company's commitment to reducing injuries and protecting your employees and company assets. Among the people who enter your front lobby are the county and state inspectors—including OSHA compliance officers—so it does pay to advertise.

d. Copies posted on company bulletin boards

Bulletin boards are another way to tell your line employees that you are serious about their safety and plan to ensure compliance.

B. Executive Management Involvement

1. Executive management provides communication to employees on accidents/injuries.

When there is a recordable injury, particularly a lost time injury, you should draft a memo for an executive member of management to sign. The memo should be addressed to all employees stating the what, where, when, how and why of the injury. This tells all employees that top management is aware of the injury/incident and has taken steps to prevent its recurrence. *The employee's name should not be used.*

2. Executive management will regularly call general employee safety meetings.

Management-employee safety meetings can be held by department, floor, job classification or shift. Meetings should be held at least quarterly. You need to give an outline of the topics to the management representative. Topics for presentation can be a comparison between where you are today and where you were a year ago with injuries, types of injuries that are still causing problems, plans that management has for the safety program, etc.

Meetings need not take longer than fifteen minutes, but they should be informative.

Give the participating member of executive management an outline of the topics to be discussed. The strategy in this meeting is that executive management (top management, plant management or whoever the employees know to have authority and clout) acts as if they already knew the information they are telling to the line employees and they are constantly monitoring the safety program. Top management will look informed and concerned.

Don't allow this to be a question-and-answer session; top management is there to communicate safety information to the employees. If questions need to be asked and answered then ask the employees to put them in writing. If your top management representative encourages questions that is fine, but the purpose of the meeting is not to put him or her in a position of being grilled by the line employees.

3. Executive management conducts monthly safety audits.

The Safety Director, the department supervisor, and members of executive management tour departments on a monthly schedule to observe how the safety program is progressing and to demonstrate participation. Don't make this an all-day affair—spend no more than one to two hours visiting scheduled departments. Break your departments into groups of three or four so that all of them can be inspected at least quarterly. I suggest you put out the schedule in January so that all departments will know when to expect a formal executive management inspection.

C. Safety Program Criterion

1. Supervisor's responsibilities and accountabilities

a. Written Procedures

You need to assure that your company has a written job description of its supervisors. Included in that job description must be at least a paragraph covering their duties and responsibilities in safety.

b. Safety Performance

Supervisors are critical to your success as the Safety Director. Your supervisors need to know that they are held accountable for their safety performance. To get their attention, remember: money talks, so talk money. An excellent method to assess accountability is through their annual evaluations. Safety performance should account for no less than 25 per cent of a supervisor's evaluation. I have found no better method of getting a supervisor's attention than pay raises and bonuses.

D. Management/Labor

1. Procedures for responding to labor suggestions are in place.

The Safety Committee is where you review the safety suggestion/hazard reports from employees. You don't have to approve all ideas for fear that you will otherwise shut off their input, but you must have the department supervisor get back to the employee who submitted the suggestion and tell the employee the decisions the committee has made. It is very important that this is done! Employees who believe that their participation is merely window dressing for management will not continue to point out areas where they feel safety can be improved. Their department supervisor is their connection to you and the committee. Therefore, the supervisor must be held responsible for reporting to the author of the suggestion the decision that was made.

II. Accident/Incident Investigation

A. Investigation

1. Accident/incident investigation priorities and procedures are in place.

All recordable injuries (any that go on the OSHA log) should be thoroughly investigated. If they are going on the OSHA log it means your injured employee will be seeking medical care outside of your facility, and this means it will appear on your insurance loss run. The responsible department supervisor will need to do a thorough accident investigation and then you, the injured employee and the department supervisor need to meet. When this recordable injury occurred, it started costing your company money and it indicates that something went wrong. It is your position to find out what went wrong and come up with corrective actions to prevent recurrence.

2. Accident/incident investigation priorities are in place for property damage.

Property damage costs money, so you need to know the what, when, where, why and how of it.

Property damage, no matter how minor, costs your company money, not only in the obvious repair or replacement of the damaged property, but also in the "down-time" when people cannot produce. Remember, this is uninsured cost—this is money right from the bottom line. What caused the incident: an unsafe act? an unsafe condition? lack of supervision? lack of training or knowledge on the part of the employee?

B. Accident/incident report routing

1. All accident/incident reports are routed through safety, personnel director, operations manager, and executive management.

Let those who need to know, know!

Don't keep your accidents/incidents a secret. It is important that you start sending the message that a safety incident is a big deal: it disrupts production, it injures people, and it costs money. Send a copy of the report to the personnel director. The injured employee may be having some difficulty in his personal life. Divorce, family illness, alcohol/drug problems, financial problems—these are areas where the personnel director can give you information, if any exists, to the circumstances in the employee's life.

The operations manager needs to know because the injured person is one of his or her employees and the injury affects production. If the injury means lost time then a replacement worker will have to be found; if medical restrictions are involved then a change in tasks may be necessary. No matter what the specifics are, the case is going to affect what takes place in that department or job task by impacting on the operations manager's ability to produce a product.

Executive management needs to know so they can also see the what, where, when, how and why. They need to know that you are on top of the event, that you have already had an investigation and have taken corrective action. They need to know that every time an accident/ incident report crosses their desk it means money is leaving the company.

2. All compensable accidents/incidents are reported to the insurance carrier within the mandated timeframe.

Your accident/incident investigation report form is an in-house form used for your internal investigation. You also need to complete a first report of injury form, which you are required to send to the insurance carrier. They in turn report it to the state. Each state has its own requirement for time frames and penalties for not reporting the injury on time (ask your claims representative or insurance agent about them). The other important reason is that if the insurance company is not notified in a timely manner, your employees' bills will not be paid. The employee will then be notified by the medical care provider requesting payment. This will only cause more problems between you and the injured employee. The message the employee will receive is, "You say you care about me, but you can't even see that the insurance carrier pays my medical bills."

C. Corrective Action

1. All accident/incident reports are analyzed for risk of recurrence.

Your meeting with the supervisor and employee will help you determine the probability of the accident/incident occurring again.

2. Accident/incident measures are developed based on elimination, substitution, engineering, training and personal protective equipment (PPE).

Ask questions, gather data and then make your recommendations. You can diminish the probability of the recurrence by making changes in the way work is done. Get as much input as possible not only from the injured employee and the supervisor, but also from other workers who do the same task. With the data before you, you can recommend engineering changes, elimination or substitution of a process, and training or retraining of an employee. Personal protection equipment may be your solution.

There can be minor solutions to minor problems. There is no sense developing an engineering solution for the problem of an employee who forgets to wear safety glasses.

In your attempt to make needed changes, you will run into the stone wall attitude that says, "What's the big deal?" Many members of management and supervisors only become concerned when a serious incident occurs. If there is not major loss of life or limb then the cause of the accident does not really seem important. Here we are talking about a matter of degrees. If the incident cost an arm or a leg then something would be done; since it didn't, let's not worry about it. You need to point out: 1.) *The exposure to the company in terms of potential and actual dollar loss* and 2.) *The violation of federal or state law, if applicable.*

Safety can be a knee jerk reactionary business—in many cases we don't close the gate until the horse is gone. Don't be reactive, be pro-active to get ahead of the problem. Use money and OSHA standards as your tools. You may not always get the changes you desire, but if you do your homework and are prepared to show your position you will be respected and listened to.

D. Investigation Training

1. Supervisors and safety committee members have been trained in accident investigation.

Training is important. Supervisors must be trained in accident investigation so they can start the process of determining what went wrong. Investigation training also forces them to take a proactive role in solving problems.

When reviewing an accident report, the safety committee must be able to tell whether an adequate job of investigation was done. Your safety committee members need to be trained so they can understand reports and be part of the search for causes and possible solutions.

Remember: top management attends your safety committee meetings and also gets a copy of the report form on the routing process. More than anyone in your company, these people need to read the report, determine the causative factors and consider the suggestions for preventing recurrence of the accident. It is also an excellent learning experience for them.

See Chapter VII for more discussion of the accident investigation.

3. Supervisors' accident reports are reviewed for completeness, adequacy of analysis and corrective measures.

The quality of a report is a litmus test both for supervisors and for you. A half-completed, sloppy accident report form tells you that the supervisor doesn't care about the incident, or did a poor job of investigation, or didn't do an investigation, or has been improperly trained. Some supervisors will test you to determine if you are really consistent in your program.

No matter what the circumstances may be, you must address the problem immediately. Draw the line and dig in! Do not allow supervisors simply to go through the motions of accident reporting. This attitude sends a message to the line employees that can severely hamper (if not destroy) your safety efforts.

You must confront the supervisor and ask why the report was not properly completed. The first line supervisor's immediate supervisor needs to be told that they are not complying with the company's policy on accident investigation. Don't let one report pass across your desk that is not properly completed. Hold your first line supervisors accountable for their accident investigations.

III.   Inspections

A.   Monthly Department Inspections

1.   All departments are inspected monthly by the department supervisor and at least one line employee.

This monthly department inspection by the supervisor and a line employee (rotated monthly) forces the supervisor to go through the department looking for unsafe acts and unsafe conditions. It gives the supervisor an extra set of eyes and a new perception when one of the employees also inspects. Because the conducting of monthly inspections is an accountability issue for the supervisor, you need to ensure that these inspections are actually being conducted.

2.   All department inspections are documented using a standard inspection form.

The key to monthly department self-inspections is the assurance that the department supervisors are really conducting them with interest and vigor. A way to ensure this is a line employee's involvement with the department supervisor. The department supervisor and one of his employees inspect the department with the intent of finding unsafe conditions that need correcting. Your supervisors and their employees are going to be watching to see if you are reading their inspection reports and following up on their discoveries.

Included in the Appendix N is an inspection form that can be used during inspections by a department supervisor and employees. The form guides them by showing them what to look for.

3. Executive management conducts a monthly inspection of scheduled departments.

It is critical to the success of your program that employees see executive management walking with you through the plant, showing interest and seriously supporting your efforts. When employees see the "big boss" come out it gives the Safety Director the influence and control needed to keep the momentum going. (Executive management inspections of departments are discussed in Chapter VIII.)

Break your departments into groups of three or four so that all of them can be inspected at least quarterly. I suggest you put out the schedule in January so that all departments will know when to expect a formal executive management inspection.

B. Safety Suggestion/Hazard Reports

1. Safety Suggestion/Hazard Reports are available to all employees.

Forms should be made available to all employees at all times to report possible hazards or make safety suggestions. Appendix U shows a sample form. I suggest you keep forms by the time clocks and in work areas, and some should always be kept at the supervisor's desk because it's important that the supervisor know the employee has a concern that needs to be addressed. In many cases the simple act of requesting the form from the supervisor will initiate corrective action from the supervisor without necessitating the completion of the form.

2. Employee Safety Suggestion/Hazard Reports are logged by the department supervisor.

This action creates accountability for the department supervisor and creates confidence in the mind of the employee that there is in fact a system in place that will follow through with his safety suggestion.

3. **Supervisors investigate Safety Suggestion/Hazard Reports and make comments on the form as to the feasibility of proposed corrective measures.**

The way to ensure that the supervisor has read the report form and discussed it with the employee is to have the supervisor make comments on the form itself. Then when you review the form, you can question the supervisor as to the why or why not of the proposed action.

4. **Safety Suggestion/Hazard Report forms are reviewed at the safety committee meeting.**

Reviewing the forms is an excellent method to determine if your employees are responding to your encouragement of their participation. It also reflects the supervisors' interest, concern and sincerity in getting their employees' input. It is here in the committee meeting that your decision can be made as to whether or not the suggestion is worth merit and should be implemented and whether or not the hazard is worth further action.

5. **Employees are notified in person by their department supervisor of the results of their Safety Suggestion/ Hazard Reports form.**

The employee needs contact from his supervisor to insure that he is being heard, and to understand that there are accountabilities in place that the supervisor will respond to. It is also important that those suggestions that are implemented get publicized for their encouragement for others to participate.

The Safety Suggestion/Hazard Report is only as good as your employees' faith in it! Your good intentions are worthless unless the employees believe that their supervisors are being held accountable for responding to their ideas. No matter how skilled you or your safety team is, one of the absolute best sources of information on how to make your company safer comes from the line employee! One of the best methods to ensure your failure is to promise your employees they will be heard and then ignore them.

IV. Safety Rules

A. General Safety Rules

1. General safety rules are in writing.

You can't play the game if you don't know the rules. General rules are just that, general in nature and specific to the point where many interpretations can't take place. General safety rules have been included in this manual in Appendix F. The rules are generic enough in nature and provide your supervisor and employees boundaries in which to operate.

2. General safety rules are reviewed at least yearly by Safety Director and supervisors.

Your safety rules need to be current and accurate for the job you intend for them. Involving your supervisors tells them you are interested in their input and want to hear their views on the relevancy of the rules they have to enforce.

3. General safety rules are posted on all plant bulletin boards.

Get the word out! Let your employees and supervisors know your expectations of them and their operating boundaries. Keep your rules visible.

4. General safety rules are used as a safety talk topic at least quarterly.

Your supervisors will be giving weekly safety talks. What better topic than a quarterly review of the general rules? Again, it is important that the rules be relevant to your operation. Quarterly is the maximum amount of time that your supervisor should go before using them as a safety talk topic.

B. Job Task Rules

1. Job task rules are in writing.

Job task rules are specific to the task the employees perform, as opposed to the aforementioned general, generic safety rules. Job task rules state specifically what the employee can and cannot do in regards to a specific task and its specific exposures. For example, an employee may be required to wear safety shoes anywhere they are in the plant—which would be a general safety rule. If, however, she operates a punch press, the specific job task rule would be to use the required safety guarding for that particular operation.

2. Job task rules are reviewed at least yearly by Safety Director and supervisors.

The review of job task rules is an area where you not only welcome the input from your supervisors, but also need it! Your supervisors know the job and probably know what safety rules are applicable to its operation. Here is a situation in which you can encourage them to demonstrate to you their safety interest and motivation. Have them make a list of safety rules that apply to certain tasks. These rules should then be reviewed at least yearly to again determine if they are relevant to the operation.

C. Safety Rule Education

1. General safety rules will be developed by employees, supervisors and Safety Director.

Included in Appendix F is a collection of generic safety rules. You can use this set as a starting point, but there can be no better method of formulating general safety rules than the involvement of your company's employees and their supervisors. Would you like to be forced to follow a rulebook that you had no say in producing?

An excellent method to use is to have each department come up with ten general rules for consideration. It gives your employees and their supervisors a sense of involvement. I keep emphasizing involvement because it is so important to the success of your program.

2. General safety rules will be covered in new employee orientation.

New employees should know the rules before they enter the work floor. They need to know your expectations of them and that you have defined boundaries they are not permitted to cross. This orientation does not have to be a threatening situation for them, but rather an informative session where they understand the company's commitment to their protection against injury.

3. A signed receipt by employees of safety orientation is filed.

After you go over the rules and the company's commitment for the employees' protection, you should have new employees make their commitment to work safely and to follow company safety rules. Have them sign a form stating that you have explained the rules and that they have read them and understand them. This will help fix in their minds the company's serious commitment to safety.

D. Safety Rule Compliance

1. Safety rule violations are covered in the company's discipline policy procedures.

Why is it that an employee will be disciplined for coming to work late, not coming to work at all, theft, and insubordination to a supervisor, but suffer no serious consequence for breaking company safety rules?

As a condition of employment, safety rule violations must be handled as seriously as any other violation. An injured employee costs you in workers' compensation, replacement of the injured worker, time lost due to accident investigation, etc. Just as you are holding your supervisors accountable, you must also hold the employees accountable. A good question to ask yourself and your supervisor when a safety rule has been violated is, "What is the consequence for this action?" The answer to that question determines the degree of response you must make.

2. The company's policy on discipline regarding safety rule violations is reviewed annually by department supervisors.

Don't panic. I didn't say supervisors had the final word on what your company's management policy is going to be, but let them have input on the consequences of the discipline they will be either handing out or recommending. They will know whether or not it works for their employees and in what cases. Use the perceptions of your supervisors as a guide in this area.

2. The policy on discipline covering safety includes:

a. Verbal Warning

Verbal warning to an employee is just that—the supervisor warns the employee verbally of a safety rule violation. However, the supervisor needs to log this in his or her daily activity book so that there is a reminder that the employee was told of the violation. Warning the employee can be done in a positive manner, such as "Jane, I know you know the safety rule covering eye protection, so please put your glasses on." Or, "Jane, do you know the safety rule on eye protection in this area?" It's important that the employee knows that he or she was observed violating a safety rule and that the company took immediate action. The word will spread quickly if your employees know your supervisors will look the other way or will enforce the policy. Consistency is the key to this warning.

b.   Written Warning

I have included samples of written warnings in the Appendices O and P. *It is important to discipline the behavior without personally criticizing the employee.* State what the employee didn't do that he or she should have been doing, or what the employee did that he or she should not have been doing. Point out the correct procedure and state that only this is acceptable. Written warnings are an excellent method of getting the attention of the employee. In your role as Safety Director, you will need to use these to the extent that you want your supervisors and employees to follow your safety rules.

c.   Suspension

If the employee still does not follow your procedures, you then need to hit home—the wallet! The employee must be convinced beyond a shadow of a doubt, your sincerity in enforcing policy. A one-or two-day suspension can do the job. The point to get across here is that your company is serious and will take the necessary action to convince the employee of it.

d.   Discharge

The concept of terminating the employment of an employee for failing to follow safety rules seems foreign to many employers. They have no problem in firing an employee for logging numerous absences, disobeying the rules of a supervisor or stealing from the company. These are indeed serious offenses, but certainly no more serious than failing to follow safe work procedures that could cost the company hundreds of thousands of dollars, and may possibly injure, maim or kill the employee and put other employees at risk. Just as with your dealings with your supervisors, you must show the employees your company's commitment to consequences. Firing an employee (particularly a long-term employee) is a difficult and painful experience—almost as difficult and painful as having an employee crippled or killed on the job because of lack of commitment and consequences. In some cases discharges must happen no matter how unpleasant the idea may seem.

E. Safety Standards

1. **Safety rules and regulations are in an employee manual.**

2. **Contents of the manual include:**

   a. Company Policy Message
   b. Safety Policy Statement
   c. Employee Responsibility
   d. Personal Protective Equipment
   e. Fire Protection/Detection
   f. Return-to-Work Policy

V. Employee Training

A. Training Requirements

1. **An audit of job tasks with high loss exposure is conducted.**

   In your company there are specific job tasks with more exposure to injury and property damage than others are. You need to determine which jobs will create the most exposure for you. Seek input from your supervisor and plant manager. After you have determined where your exposures are, you then need to ensure that necessary safety precautions are taken for these specific tasks.

2. **All training is compared to federal and state mandates.**

   This policy item tells your employees that you are interested in their welfare and ensures that the company is in compliance with the law.

All training is important, but keep in mind that there are federal and state regulations requiring annual training—arrange to fulfill these requirements first! Have your insurance loss control representative advise you as to what required training you must provide. I have included in the References a list of companies that offer videos, booklets, etc. that will provide the type of information your employees need to have.

B. New Employee Orientation

1. A safety orientation is provided to all new employees.

It is vital to your program that you start new employees with the perception that safety is as important as production. Take at least an hour with your new employees to go over all areas of your safety program, letting them know how important safety is not only to you, but also to them.

This is the opportunity for you to state what the company's expectations of the employees are and what the consequences of their actions will be. From the very start of your relationship with the new employees, demonstrate the attitude that you and the company are very serious about preventing injuries.

2. A three-month follow-up session for new employees is conducted.

When you initially have the new employee for the safety orientation, you must be realistic about how much the employee will retain. A new employee is going to be overwhelmed with so much new data that there will be much this person is going to miss. Three months later, you need to determine how much she or he remembers and uses. More importantly, you can tell from the new employee's attitude at the follow-up if his supervisor is reinforcing what you said.

This three-month orientation is simply an audit of the employee and your ability to convince him of your company's commitment. Ask questions and try to determine if the employee feels motivated to work safely after settling in. If not, why not? The three-month audit tells the employee that you haven't forgotten your commitment to him and that you haven't forgotten your expectations of this person's job performance.

C. Job Instruction

1. A written policy requires supervisors to give job instruction to each new employee.

Never assume that your employee knows how to work safely, or for that matter never assume that he or she even thinks safety is really a part of a job. Ask yourself where your employees have been trained to work safely. If they were not in the military service or a high school shop class,—where safety is usually at least partly taught— there is a great chance that they have never been properly safety trained. Industry is slowly learning that if you want your employees to work safely, you must train them to work safely.

D. Supervisor Contacts

1. A written standard requires supervisors to "safety tip" their employees on a daily basis.

The supervisor needs to be actively involved with their employees on a daily basis regarding safety. Your supervisor should visit each employee at least daily—if not more often—and use the word "safe," "safely," or "safety" in a sentence. As simplistic as this "safety tip" method may sound, it works! It forces the supervisor to make rounds and to sell the safety message. Safety tipping is not time-consuming and provides benefits well worth the effort.

E. Safety Meetings

1. A written policy requires supervisors to give weekly safety talks.

Get ready, Safety Director. This may be the area where your supervisors will howl and scream. The number one fear of many people is public speaking. Your supervisors are no different from other people. Nevertheless, your policy requires them to stand up in front of their employees and give a safety talk.

It is not the particular safety message given at the talk that is so crucial, but the fact that the message is given. Start your supervisors out with purchased safety talks that can be read aloud. Later on you can give them the leeway to use their own topics or the ones you choose.

The important point is that supervisors do it on a weekly basis, preferably on Monday, and that they are held accountable for it. You should ask your employees what their safety talk was about and how they received the message. This will ensure compliance by the supervisors.

2. Records of safety talks and safety meetings are maintained.

Require all employees to sign attendance sheets when they take safety training. It gives them the impression that you are serious about this effort and will hold them accountable for following the training given. It also tells OSHA that you are serious about mandated training and that you can provide proof that such training was given.

Require that the attendance sheets be turned in to you weekly.

F. Safety Standards

1. Safety rules and regulations are in an employee manual.

Give your employees a handbook that outlines your commitment to the safety program and your expectations of them. This can be open-ended in that there will not be rules for each specific task, but a general overview on your position protecting the worker and company from loss. In Appendix F I have included a copy of an employee manual for your use.

2. Contents of the manual include:

a. Company Policy Message

This letter should be a joint message signed by the company president, the safety committee and the union if applicable. A sample of a policy message is found in Appendix C.

b. Safety Policy Statement

The safety policy statement is separate from your policy message in that it is the policy of the company and represents its positions on the subject.

c. Employee Responsibility

The employee responsibility statement gives the company's expectations of the employees. Again, it is open-ended but does convey the message that the company holds the employees accountable and responsible for their actions.

d. Personal Protective Equipment

The equipment statement gives an overview of what the company's position is on the wearing of personal protective equipment and the enforcement of its use. It does not need to be specific to each job task, but rather a general overview of the topic.

e. Fire Protection/Detection

The fire protection and detection statement gives the company's position regarding the protection of its employees in case of fire and regarding fire prevention.

f. Return-to-Work Policy

This statement tells the employee that you have a policy that will return the injured employee to work as soon as possible.

This reinforces the message that the position of the company is for accountability.

VI. Personal Protective Equipment (PPE)

A. Personal Protective Equipment Standards

1. Standards for specific jobs are in writing.

Personal Protective Equipment Standards (PPE) ensure there is a barrier between your employee and the hazards of his job. Not all jobs will require safety glasses, respirators, gloves or other PPE, but some do.

Those jobs that do need specific protection should have a policy stating the job or task and the type of PPE that is mandatory. The supervisor needs to enforce this policy, and there must be consequences for failing to follow this policy.

2. Supervisors and employees are trained in the need for PPE.

It's one thing to tell your employees they must wear PPE and another to convince them that it is for their protection. Training is the way to get the employees to understand the need for safety glasses, shoes, or other PPE. I suggest you first train your supervisors through videos and booklets and then bring your employees into the PPE training. Many safety suppliers will provide free informational material to help you in your training.

Unfortunately, you can give your employees as much training as possible and yet you are still going to get employees who will give every reason in the world why it won't work, is not necessary and is an invasion of privacy, etc. The key here is for the supervisor to hold the line and set the example.

B. Compliance of PPE Standards

1. Written disciplinary action for PPE violation is in effect.

If your company has a written policy on PPE but your employees won't follow it and your supervisors don't enforce it, then there must be consequences. The company is committed to preventing injuries and PPE helps you do that. If employees refuse or only mildly comply and the supervisors look the other way, then there must be disciplinary action.

VII. Medical Considerations

A. Professional Medical Consultation

1. The company medical consultant treats injured employees.

Talk to your agent and your claims representative to be sure about your state's law regarding treating physicians. Most states permit the company to have the injured employee see the company medical consultant.

2. The company medical consultant attends monthly claims review.

Your claims review meeting is the time when diagnosis and prognosis of the injured employee will be discussed. You need your company medical consultant there to give you the inside picture of what is going on with this employee—what he can do and what he can't. Also, if the injured employee has another treating physician, you need to have your doctor look at the restrictions and determine if they are realistic, or whether your company physician should call the treating physician to ask for clarification and to say that there is indeed modified duty available.

3. **An employee assistance program is in place to assist these employees with personal problems.**

Everyone at some time or another in life hits choppy water: divorce, alcohol, drugs, financial problems, etc. Your employees need somewhere to go for problem resolution and guidance. Your employees bring those problems to work and the problems manifest themselves in a person's work activity. Although your supervisor and the personnel department can be of assistance in this area, your company needs to have a professional employee assistance program (EAP) in place.

C. First Aid Training and Supplies

1. Supervisors receive annual first aid and CPR training.

Remember that an employee who must leave your company to receive outside medical care will also be recorded on your OSHA log (although there are certain first aid injuries that OSHA exempts from the log). Once the employee receives outside medical care, your insurance company will be paying for the medical care and recording the bill. In many cases, the difference between a first aid injury and a recordable injury is the difference between your supervisor treating the injured employee or simply telling him or her to go to the doctor.

When employees are injured, it is important that they know they are going to receive immediate attention by someone who has been trained in the basics. There are many firms that provide two-hour first aid courses and a four-hour CPR course. Your supervisor does not have to be a paramedic, but should be able to treat cuts, lacerations, and minor injuries. The rule of thumb for your supervisors should be, "When in doubt, send them out."

2. Each department has a fully stocked first aid cabinet.

In order to provide proper and immediate medical care to your employees, there should be a first aid cabinet in each department. The cabinet should be stocked with the type of first aid equipment (bandages, eye wash solution, etc.) that is relevant to the type of work that your employees are performing. There are many firms that can inventory your first aid supplies and make recommendations.

It is important that the first aid cabinet is inventoried monthly by your department supervisor and that needed items ordered.

3. **Employees are required to report all injuries to their supervisors for treatment and/or referral to the company medical consultant.**

Your supervisors must insist that all injuries are reported to them. This procedure serves two purposes. First, it alerts a supervisor that an incident has occurred in the department. Because the incident caused an injury, the employee's immediate supervisor needs to find out why.

Injuries don't just happen. A first aid scratch may seem minor, but it represents something that went wrong. *First aid injuries are a warning bell you should be listening for—they tell you the next step is a recordable lost time injury, or even a fatality.* Don't assume that if "it's only a scratch" it is therefore insignificant. Don't be reactive and only respond to lost time and deaths after the fact. Instead, listen when the warning bell goes off. Be proactive not reactive—find out what went wrong and prevent another occurrence.

Second, the policy that requires reporting of injuries prevents the injured employee from treating himself or, even worse, simply ignoring the injury.

Don't let employees give themselves first aid. You can't be sure they have the know-how. The procedures they use and the medications they choose from the first aid box might be the wrong treatment. Supervisors that have been trained in first aid should be the ones designated to provide medical care. Never let employees treat themselves—never!

VIII. Navigate Program

A. A Written Policy on Return-to-Work Is In Place.

Your employees need to know that when they are injured they are not going to be forgotten. They need to know that they are still considered part of the company, and that the company will keep in touch with them. Too many companies have the attitude, "Out of sight, out of mind." The insurance carrier may be paying the medical bills, but injured employees know that they were hurt working for your company.

1. The return-to-work policy is communicated to all employees.

Communicate by company policy, memo, or group/department meetings. Let your employees know they are going to be monitored if they are off work—the company will stay in touch.

2. Supervisors are trained in the return-to-work policy.

Your supervisors need to understand why the company brings injured employees back to work. Through training, you show them the cost of leaving an employee out on workers' compensation and the impact that a workers' compensation claim has on the bottom line of the company. In addition, they need to understand the injured employee's restrictions and limitations and practice being sensitive to them. They also need to hold the company line when there are complaints from other employees who claim they are doing more work than their modified duty peers are. Many return-to-work programs have failed simply because supervisors did not support them.

3. Injured employees who return to work under a modified duty status will be assigned work based on functional capacities.

Included in Appendix R is a medical capabilities form. It not only tells the company what the employee can't do, but it also tells you what the employee can do. Placing an employee within restrictions is difficult, so many companies throw up their hands, groan, and accept the loss. When you find that happening in your company, find out from your claims representative how much money it will cost the company to take the easy way out, versus the costs of bringing the employee back. Also ask what impact this will have on your future premiums. Then it is time for you to remember "money talks, so talk money." Use your knowledge to convince management and supervisors of the effectiveness of modified duty. (And when you lose a round, remind yourself, "If it were easy, everyone would be doing it.")

B. Navigate Coordinator

1. A Navigate coordinator is appointed.

The Navigate coordinator could be an individual from the personnel department, or even you. This person is responsible for maintaining contact with off-work employees and needs to be sensitive to their concerns.

2. The Navigate coordinator makes weekly contact with lost time injured employees.

One of the biggest concerns the off-work injured employee has is that "No one at my company cares about me. They used me when I was working for them, but now that I'm not useful to them I never hear a word."

Unfortunately for the employer, the friendly voice most injured employees hear on the other end of the phone is not someone from the employee's company, but rather a stranger from an insurance company. I believe that if an employee were hurt bowling or boating, there would typically be more phone contact from the employer than if the employee were hurt on the job. Why? It goes back to the assumption made by the injured employee's company that "The insurance company is handling it." The employee will start playing psychological mind games when this occurs—"They really don't care," "My boss never liked me," "It's their way of getting rid of me," etc. Little wonder there are so many workers' compensation attorneys whose advertisements claim that they will show concern, interest and communication. Apparently they operate thriving businesses by doing so!

Be a friend to these injured employees—you need them!

Your Navigate coordinator can communicate concern and interest simply by using two devices: the phone and the mail. A weekly phone call can ask questions such as "How are you?" or, "Are you getting your checks from the insurance company on time?" The caller should be conversational, saying, "We miss you, your job is here when you come back, take it slow, don't push yourself," etc. This phone call should not be conducted in such a way as to be interpreted by the off-work employee as the company's device to find out if he or she is really at home. Make it clear that it's the company's way of staying in touch and showing that it cares.

Buy a box of generic get-well cards to have on hand. When an injured employee is off work, periodically send a card signed by the supervisors, fellow employees, and yes, even the company president. Let the employee know that you are sincerely interested in his or her well being, recovery and return-to-work. The greeting card technique really works.

3. The Navigate coordinator contacts the insurance company claims representative for updates on lost time injury cases.

In dealing with the injured employee, the Navigate coordinator represents the sincere concern and interest of the company. But the Navigate coordinator also needs to keep an eye on the bottom line— What is this injury costing the company? What is the medical prognosis in this case? How long is the treatment and recovery time going to run? What activity is the claims representative doing to get this injured employee back to work?

The Navigate coordinator keeps the insurance company's feet to the fire. It is acceptable for this person to convey the attitude that your company thinks it is the only client the insurance claims representative has. In seven and one-half years working with an insurance company, I learned that the axiom "The squeaky wheel gets the oil" is oh so true. Your company must keep asking the insurance carrier to provide you with the information you need. If no information comes forth, call your agent!

IX. Safety Promotion/Information

A. Bulletin Boards

1. Bulletin boards will be used to display safety posters, and these posters will be changed weekly.

Posters can spread the concept of safety. Through their use of humor and drama, they constantly remind employees of your interest and commitment to the program. The vendor I have used is Clement Communications because their posters have eye appeal and humor.

Individual safety posters that are left posted for an extended period of time indicate to your employees that you lack consistency and commitment. For safety posters to be effective, you must change them weekly. (Don't be surprised if your employees write on them—it means they are seeing the posters.)

2. Bulletin boards are used to display notices, memos and other relevant safety-related information.

If a memo is confidential and/or sensitive, sending it through normal routing channels to the supervisors is fine. But if you are notifying your supervisors that there will be safety training on a subject, don't keep the announcement a secret.

You need high exposure for your safety efforts. Keep your employees involved and interested by posting safety information, and don't depend on your supervisors to get the word out promptly. By posting memos and notices, you encourage your supervisors to be accountable to their employees and to you.

X. Incentive Awards

A. Department Recognition

1. Those departments with no recordable injuries will be recognized.

There is a big difference between bribing employees and giving them an incentive. Nothing can turn off the momentum of a safety program quicker than to approach top management to "pay employees to work safely." When worded that way, it is easy to understand why that concept can cause problems for an employer.

There is competition in almost every aspect of our lives and certainly work is no exception. Peer pressure, when properly directed and monitored, can stimulate more involvement from your employees and supervisors. A technique I have used successfully is to develop a criterion that measures one department against another. There are three measurement devices: **1**. *No recordable injuries for the past month*; **2**. *Compliance by all (including department supervisors) with company rules and policies;* and **3**. *Overall general housekeeping in the department.*

Members of the safety committee go through the plant making observations at the end of the month, a vote is taken, and the winning department has pizza delivered to them for their lunch break. It is not particularly expensive, but it does send a clear message that there are rewards for getting on the team.

B. Individual Recognition

1. Individuals who have gone a quarter without a recordable injury will be rewarded.

It is important to reward employees on an individual basis for working safely. Look at your loss runs to see what just one recordable injury can cost the company. Keep that fact in mind when you plan how to sell the incentive concept to your top management.

It is a program I have used successfully. I prefer the quarterly reward program because it allows the employees to "bank on themselves" and because yearly gifts don't motivate employees who already have a recordable injury, and eliminates them from the program entirely.

SAFETY PROGRAM: Condensed Outline

I. MANAGEMENT

A. Safety Policy Statement

1. The company has a general policy that confirms and proclaims its commitment to accident prevention.
2. The policy is signed by the chief executive officer.
3. The policy statement is distributed via:
   a. A copy sent to each department manager.
   b. A copy sent to each first line supervisor.
   c. An enlarged copy posted in the main lobby.
   d. Copies posted on company bulletin boards.

B. Executive Management Involvement

1. Executive management provides communication to employees on safety.
2. Executive management holds general employee safety meetings.
3. Executive management conducts monthly safety audits.

C. Safety Program Criterion

1. Supervisor's responsibilities and accountabilities.

D. Management/Labor

1. Procedures for receiving and reviewing recommendations from labor are in place.

2.   Procedures for responding to labor suggestions are in place.

II.   ACCIDENT/INCIDENT INVESTIGATION

A.   Investigation

1.   Accident/incident investigation priorities and procedures are in place.
2.   Accident/incident investigation priorities are in place for property damage.

B.   Accident/Incident Report Routing

1.   All accident/incident reports are routed through safety, personnel director, operations manager, and executive management.
2.   All compensable accidents/incidents are reported to the insurance carrier within the mandated time frame.

C.   Corrective Action

1.   All accident/incident reports are analyzed for risk of recurrence.
2.   Accident/incident measures are developed based on elimination, substitution, engineering, training and PPE.

D.   Investigation Training

1.   Supervisors and safety committee members have been trained in accident investigation.
2.   Supervisors' accident reports are reviewed for completeness, adequacy of analysis and corrective measures.

III.   **INSPECTIONS**

A.   **Monthly Department Inspections**

1.   All departments are inspected monthly by the department supervisor and at least one line employee.

2.   Executive management conducts a monthly inspection of scheduled departments.

B.   **Safety Suggestion/Hazard Reports**

1.   Safety Suggestion/Hazard Reports are available to all employees.

2.   Employee Safety Suggestion/Hazard Reports are logged by the supervisor.

3.   Supervisors investigate Safety Suggestion/Hazard Reports and make comments on the form as to the feasibility of proposed corrective measure.

4.   Safety Suggestion/Hazard Report forms are reviewed at the safety committee meeting.

5.   Employees are notified in person by their department supervisor of the results of their Safety Suggestion/Hazard Report form.

IV.   **Safety Rules**

A.   **General Safety Rules**

1.   General safety rules are in writing.

2.   General safety rules are reviewed at least yearly by the Safety Director and supervisors.

3.   General safety rules are posted on all plant bulletin boards.

4.   General safety rules are used as a safety talk topic at least quarterly.

B. Job Task Rules

1. Job task rules are in writing.
2. Job task rules are reviewed at least annually by Safety Director and supervisors.

C. Safety Rule Education

1. General safety rules are developed by employees, supervisors and Safety Director.
2. General safety rules are covered in new employee orientation.
3. Receipts are signed by all employees confirming that they have received a safety orientation.

D. Safety Rule Compliance

1. Safety rule violations are covered in the company's discipline policy procedures.
2. The company's policy on discipline regarding safety rule violations is reviewed annually by department supervisors.
3. The policy on discipline for safety violations includes:
    a. Verbal warning
    b. Written warning
    c. Suspension
    d. Discharge

E. Safety Standards

1. Safety rules and regulations are in an employee manual.
2. Contents of the manual include:
    a. Company Policy Message
    b. Safety Policy Statement
    c. Employee Responsibility
    d. Personal Protective Equipment
    e. Fire Protection/Detection

f.   Return-to-Work Policy

V.   Employee Training

A.   Training Requirements

1. An audit of job tasks with high loss exposure is conducted.
2. All training is compared to federal and state mandates.

B.   New Employee Orientation

1. A safety orientation is provided to all new employees.
2. A three-month follow-up session for new employees is conducted.

C.   Job Instruction

1. A written policy requires supervisors to give job instructions to each new employee.

D.   Supervisor Contacts

1. A written standard requires supervisors to safely tip their employees on a daily basis.

E.   Safety Meetings

1. A written standard requires supervisors to give weekly safety talks.
2. Records of safety talks and safety meetings are maintained.

F.   Safety Standards

1. Safety rules and regulations are in an employee manual.

2. Contents of the manual include:
   a. Company policy message
   b. Safety policy statement
   c. Employee responsibility
   d. Personal protection equipment
   e. Fire protection/detection
   f. Return-to-work policy

VI. Personal Protective Equipment

A. Personal Protective Equipment Standards (PPE)

1. Standards for specific jobs are in writing.
2. Supervisors and employees are trained in the need for PPE.

B. Compliance of PPE Standards

1. Written disciplinary action for PPE violations is in effect.

VII. Medical Considerations

A. Professional Medical Consultation

1. The company medical consultant treats injured employees.
2. The company medical consultant attends monthly claims review.

B. Employee Assistance Program

1. An employee assistance program (EAP) is in place to assist those employees with personal problems.

C. First Aid Training and Supplies

1. Supervisors receive annual first aid and CPR training.

2.   Each department has a fully stocked first aid cabinet.

3.   Employees are required to report all injuries to their supervisors for treatment and/or referral to the company medical consultant.

VIII.   Navigate Program

A.   A Written Policy on Return-to-Work Is in Place.

1.   The return-to-work policy is communicated to all employees.

2.   Supervisors are trained in the return-to-work policy.

3.   Return-to-work of injured employees to modified duty status is based on functional capacities.

B.   Navigate Coordinator

1.   A Navigate coordinator is appointed.

2.   The Navigate coordinator makes weekly contact with lost time injured employees.

3.   The Navigate coordinator interfaces with the insurance company claims representative to be updated on lost time injury cases.

IX.   Safety Promotion/Information

A.   Bulletin Boards

1.   Bulletin boards are used to display safety posters, changed weekly.

2.   Bulletin boards are used to display notices, memos and other relevant safety-related information.

X.   Incentive Awards

A.   Department Recognition

1.   Those departments with no recordable injuries for a month are recognized.

B.   Individual Recognition

1.   Those individuals who have gone a quarter without a recorded injury are rewarded.

Appendices

Safety Director Job Description

Position Title: Safety Director   **Date**:
Division: **Approvals**:
Accountable To:

Primary Objective of Position:

To establish and promote safe working conditions and practices, to develop accident prevention and loss control programs, and to design procedures for the evaluation of these programs.

Major Areas of Accountability:

1. Coordinate safety activities throughout the company.
2. Maintain and analyze all accident reports for thoroughness and clarity.
3. Determine and coordinate safety education for company employees.
4. Coordinate activities to maintain interest of company employees using bulletin boards, posters, and other informational transmitting methods as needed.
5. Coordinate accident investigations and near miss reports.
6. Coordinate all internal safety inspections.
7. Ensure current knowledge of OSHA standards.
8. Issue reports showing accident/incident patterns and trends.
9. Serve as the company's contact with your insurance carrier.
10. Serve as the primary contact with OSHA compliance officers.
11. Maintain current recordkeeping and training records as mandated by OSHA.
12. Keep the company president briefed and informed on the status of the safety program.
13. Perform other duties as required.

Principles of Safety

Ten Principles of Safety

1. All injuries and occupational illness can be prevented.

2. Management is directly responsible for prevention of injury and illness.

3. Safety is a condition of employment.

4. Training is essential for a safe workplace.

5. Safety audits are a must.

6. All deficiencies must be corrected promptly.

7. It is essential that all unsafe practices, accidents and Injuries be investigated.

8. It is good bu$ine$$ to prevent injuries and illnesses.

9. People are the most important element.

10. Safety off-the-job is as important as safety on-the-job.

Company Safety Policy Statement

Safety Policy Statement

Our company has a vital interest in the prevention of losses due to accidents. This is necessary because it involves the safety and the well being of all employees, customers, and the public that we serve. Injuries and accidents can be prevented.

The creation and maintenance of a safe working environment requires the interest, cooperation, and dedication of every employee in our organization. A safe work environment only exists when all employees observe safety procedures as an integral part of every work procedure.

All levels of management will make accidents and loss prevention a matter of main concern. Safety ranks right along with efficient production of our quality products for the customers that we serve.

We acknowledge our obligation to provide our employees with a place of employment that is free of recognized hazards that are likely to cause death or serious physical injuries. The responsibility for the safe operation of each department rests with the supervisor in charge of that area.

It is the responsibility of each employee to accept and comply with all safety and health standards and work rules, regulations, and instructions that are applicable to their own actions and conduct.

_____
President

OSHA General Duty Clause Policy Statement

Safety Policy Statement

Subject: OSHA GENERAL DUTY CLAUSE

General

1.   The Occupational and Safety Health Act General Duty Clause states:

"Each Employer shall furnish to each of his employees, employment and a place of employment which are free from recognized hazards that are causing or likely to cause death or serious physical harm to its employees. Each employer shall comply with the Occupational Safety and Health standards promulgated under this act. Each employee shall comply with occupational safety and health standards and all rules, regulations and orders pursuant to this act which are applicable to his own actions and conduct."

Company Responsibilities

1.   The Company will abide by the Occupational Safety and Health Act.

Employee Responsibilities

1.   Employees are responsible for reporting safety problems, including unsafe acts and/or unsafe conditions.

2.   Employees are responsible for abiding by all OSHA and company safety and health rules and practices.

WORKING SAFELY IS A CONDITION OF EMPLOYMENT

Annual Training Policy Statement

Safety Policy Statement

Subject: ANNUAL TRAINING

General

1.   Our Company provides formal annual training to all plant personnel per OSHA mandated requirements.

Training Requirements

1.   Annual training for all employees who are exposed to noise at or above an 8-hour time-weighted average of 85 decibels.   OSHA 1910.95 (K) (L).

2.   Annual training to familiarize employees with general principles of fire extinguisher use.   OSHA 1910.157 (G).

3.   Annual training for all employees to provide information and training on hazardous chemicals in their area.   "Right-to-know".   Hazard Communication 1910.1200 (H), (I), (Z).

WORKING SAFELY IS A CONDITION OF EMPLOYMENT

General Safety Rulebook

The management of Acme Industries holds in high regard the safety, welfare, and health of its employees. We believe that production is not so urgent that we cannot take time to do our work safely. In recognition of this, and in the interest of modern management practice, we will constantly work toward:

The maintenance of safe and healthful working conditions.

Consistent adherence to proper operations, practices, and procedures designed to prevent injury and illness.

Conscientious observance of all federal, state, and company safety regulations.

President_____

Operations Manager_____

Safety Director_____

Acme Industries makes every effort to provide a work place free of recognized hazards. These general rules have been established for your guidance. In your own interest, you must familiarize yourself with every rule set forth.

Because of the variation in the work of departments, it is impossible to include in this publication all the rules governing safety of operations. Rules that apply to specific operations or processes will be brought to your attention by your supervisor.

At the end of this booklet is a tear-out sheet which you must sign and date. This sheet will become a permanent part of your personnel records. This is being done in your best interest and we are convinced that it will make your employment with the company injury free.

Safety Committee

Know Your Job

If you are unsure about the safe and proper way of doing your job, ask your supervisor. Working safely is a condition of employment. Your safety and the safety of everyone in the department is vitally important to all Acme supervisors.

Unsafe Conditions

If you see any ways to improve safety conditions in your area, report them to your supervisor immediately. If you see any unsafe conditions, report them immediately. Your supervisor will take steps to correct them. If you feel that improper or insufficient action has been taken, you can contact any member of the Safety Committee for further action.

Complete Instructions

Before attempting to operate any machine or equipment, know the safe way to do the job. Unauthorized use of equipment or machinery is prohibited. Adjustment and repair of machinery is to be done by authorized personnel only.

Housekeeping

The most important personal contribution to safety is a clean work area. Always keep your place in a neat and orderly manner.

Safety Shoes

Safety shoes are required in all operating areas of the plant. The company offers a subsidy to help defer the cost of safety shoes.

Eye Protection

All employees must wear safety glasses. Non-prescription glasses are available from your supervisor. If prescription glasses have to be ordered, you will be required to wear your regular glasses with goggles until your prescription glasses arrive. The company offers a subsidy to help defer the cost of safety glasses.

Gloves

Gloves are required in some areas of the plant. Your individual supervisor will instruct you on the use of gloves. Gloves must not be worn when operating or working in close proximity to rotating machinery.

Fork Trucks

Only trained and authorized operators shall be permitted to operate powered fork trucks. Employees are prohibited from riding on the forks of trucks.

Safe Tools

A good mechanic uses tools properly and keeps them in good repair. It is your responsibility to properly exchange tools that become unsafe. Many operations require specially designed tools. Do not use makeshift tools. See your supervisor for the proper tools for the job.

Guards

Guards on machines are there for your protection. If a guard has been removed be sure it is replaced before operating the machine. If you think an additional guard is needed, report this to your supervisor.

Electrical Repairs

Electrical repairs must be made by an authorized electrician. Service personnel must lock out the master disconnect switch before working on any machine. Report any necessary electrical repairs to your supervisor immediately.

Air Hose

Always use an air hose with caution. Do not direct the air nozzle towards an aisle, another person or yourself. No air hose can exceed 30 P.S.I. If unsure of the P.S.I. ask your supervisor to test it before using.

Miscellaneous

Never stand a pallet on edge or use a pallet as a platform.

Parking lot safety is important—drive with care. Observe the proper one-way traffic and posted speed limits.

Observe the **NO SMOKING** areas.

Use of alcoholic beverages and/or narcotics is strictly prohibited.

Hearing protection is required in selected areas. See your supervisor for hearing protection equipment.

Learn to recognize unsafe conditions and report them to your supervisor immediately.

Fire Rules

Remain calm under all circumstances. Exercise good judgment and consideration for the welfare of others.

By remaining at your station, the emergency operation necessitated by a fire will be limited to the employees in the immediate vicinity of the fire and the authorized fire brigade; thus any confusion will be eliminated.

All employees are to be familiar with emergency exits and emergency equipment. **DO NOT BLOCK EXITS** or emergency equipment.

Fire extinguishers are located throughout the plant. All extinguishers are labeled with operating instructions and type of fire designations.

Familiarize yourself with the locations of fire extinguishers near you work station. **DO NOT BLOCK FIRE EXTINGUISHERS.**

All flammable and combustible liquids must be kept in closed, approved containers.

Containers must be labeled as to contents.

Rags saturated with flammables must be kept in closed, approved containers.

Never pour oil, flammable liquids, or any other chemical into any sewer or drain.

Smoking is prohibited in restricted areas.

In case of fire or emergency, dial the operator, give your name, exact location of the fire and/or emergency, and any other pertinent information.

Evacuation of Building

If the fire and/or emergency cannot be handled by the fire brigade, the operator will instruct the personnel to leave the building. The operator will then call the fire department.

All personnel, other than the fire brigade, will walk (**NO RUNNING**) to the nearest exit and evacuate the building to allow the Fire Department free access.

All individuals assigned to fire brigade will check their area to ensure that everyone is safely out of the building before exiting himself or herself.

In Conclusion

These are your rules for safe operation. Years of experience have shown them to be the safest way to perform your daily work.

The key to accident prevention lies with the individual, not only in the observance of these rules, but more importantly, in the acceptance of a personal responsibility for being mentally alert, and using good judgment and common sense at all times.

The Company provides you with training, up-to-date tools and equipment, and the best safety rules that can be devised. But these are only a secondary consideration; the primary responsibility belongs to YOU. You owe it to yourself and to your family to work safely.

Remember, working safely is a condition of employment.

I, _____, have read and understand this booklet and realize that working safely is a condition of employment.

Date _____

Failure to comply with the safety guidelines described in this booklet can result in disciplinary action, up to and including termination.

Employee Injury Monthly Report Card

DATE:_____

	YEAR TO DATE	
	LAST YEAR	THIS YEAR
TOTAL RECORDABLE INJURIES	_____	_____
TOTAL DAYS LOST	_____	_____
RECORDABLE INJURIES WITH DAYS LOST	_____	_____
RECORDABLE INJURIES BY TYPE:		
MISC	_____	_____
EYE	_____	_____
STRUCK AGAINST	_____	_____
STRUCK BY	_____	_____
FALL, SLIP, TRIP	_____	_____
BODY MECHANICS	_____	_____
CAUGHT IN BETWEEN	_____	_____
LAC, CUT, TEAR, PUNCT	_____	_____
CONTACT WITH TEMP EXTREMES	_____	_____
LOST TIME RECORDABLE INJURIES BY TYPE:		
MISC.	_____	_____
EYE	_____	_____
STRUCK AGAINST	_____	_____
STRUCK BY	_____	_____
FALL, SLIP, TRIP	_____	_____
BODY MECHANICS	_____	_____
CAUGHT IN BETWEEN	_____	_____
LAC, CUT, TEAR, PUNCT	_____	_____
CONTACT WITH TEMP EXTREMES	_____	_____

Year-to-Date Workers' Comp Injuries by Department

Number of Recordable Injuries Through:_____

Department Number	Injuries	Dollar Cost of Injuries to Company
1		
2		
3		
4		
5		
6		
7		
8		
9		
10		
TOTAL		

Periodic Loss Report Form

TO: Your Boss   <u>SAMPLE</u>

FROM: You, the Safety Director

DATE:

SUBJECT: Loss Report W/C Claims Processed through 5/31/98

Your Company	TOTALClaims including Reserves
1993	804,222
1994	420-875 -48%
1995	858,496 +104%
1996	198,950 - 77%
1997	127,026 -36%
1998	- 84%

SAMPLE CALCULATION: FOR 1994

(804,222-420,875)   =383,347
(383,347/804,222)   =0.476   = 48%

Note: In the preceding sample calculation, 1994 was compared to the preceding year. 1994 was compared to the baseline year (1993). You can do either. In our example, the drastic rise in 1995 was due to 2 unusual accidents, and not reflective of the improvement of overall company safety.

177

TO:

FROM:

DATE:

SUBJECT: Loss Report W/C Claims Processed through

Your Company	TOTALClaims including Reserves

Monthly Cumulative Incident Rate by Department

DATE: _____

Department	Hours Worked	Year-to-Date Incidents	Year-to-date Incident Rate

TOTAL   TOTAL   TOTAL
YEAR TO DATE

_____   _____   _____

Formula: (# incidents X 200,000)/Total Hrs Worked=Incident Rate

National Yearly Incident Rate for our SIC Code _____

Year-end Department Incident Rate

National average for our SIC code is 19.0
Year-end company incident rate is 7.0

Department	Incident Rate
11	0
12	0
17	0
18	0
20	0
22	0
23	0
27	0
28	0
29	0
24	2.2
19	3.1
13	3.6
16	9.3
10	9.6
26	10.9
21	13.5
25	21.5
30	23.4
14	24.4
15	26.4

YEAR END DEPARTMENT INCIDENT RATE

National average for our SIC code is _____

Year-end company incident rate is _____

Department	Incident Rate

Claims Status Report

DATE: _____

NAME: _____

DOI: _____

TYPE OF INJURY: _____

LAST MEDICAL APPT: _____

RESULTS:

_____
_____
_____
_____
_____
_____
_____
_____
_____

NEXT MEDICAL APPT: _____

REGULAR JOB: _____

MODIFIED JOB: _____

ORC: _____

TOTAL COST TO DATE: _____

PLANS/COMMENTS: _____

Maintenance Work Order

MAINTENANCE WORK DATE:_____REQUESTED BY:_____

DEPT.:_____ SAFETY RELATED:_____

LOCATION:_____ *SAFETY PRIORITY: _____

STATE SPECIFICALLY WHAT WORK YOU WISH TO BE PERFORMED:

_____
_____
_____
_____
_____
_____
_____
_____
_____
_____
_____
_____
_____
_____

* Must be initialed by Safety Director

Department Inspection Report Form

Area Inspected_____

Date:_____   Inspected by: Supervisor:_____
                                        Employee: _____

Use these categories as a guide for listing items on this inspection:

__ Ventilation   __ Vehicles

__ Floors, doors, stairs   __ Toxic material storage, labels

__ Flammable liquid, gas labels, storage __ Containers

__ Aisles, work surfaces   __ Equipment, guarding

__ Lighting   __ Personal protective equipment

__ Electrical wiring, cords   __ First aid

__ Exits, alarms, emergency lighting   __ Smoking

__ Fire protection equipment   __ Housekeeping

__ Hand and power tools   __ Warning signs, labels

__ Ladders, scaffolds   __ Other

LIST HAZARDS AND CORRECTIVE ACTION TAKEN:

_____
_____
_____
_____

Employee Safety Discipline Letter #1

Attn: Robert Smith, Department 4

Dear Robert:

On May 4, 1998, you met with Jim Wilson, your union representative; Larry Wilcox, plant director; Jerry Thomas, department supervisor; and me to discuss your job performance on May 21, 1998.

While operating a band cutter you lacerated your thumb and index finger to the extent that over six sutures were required. The accident investigation revealed that you were not wearing gloves and did not report the injury to your department supervisor.

It was discussed at the meeting that you know that the wearing of gloves while operating the band cutter is a company work rule. It was also agreed at the meeting that all injuries, no matter how minor, are to be reported to your department supervisor immediately.

Your responsibility as an employee of the company requires you to follow work safety rules, which are a condition of your employment. This is the second incident within the past six months when you have violated a safety rule resulting in injury to yourself.

The company is mandated by state and federal law to protect you and all other employees from injuries committed by unsafe acts or unsafe conditions. You acknowledged in the meeting your interest in safety and I'm confident that your interest is genuine. However, safety on the plant floor must have the same priority to you as quality and production.

This letter will remind you that you must concentrate on safety while performing your work. You understand that further corrective action will be taken for any further carelessness or negligence. A copy of this letter will be placed in your personnel file.

Yours truly,

Nancy Smith
Personnel Director

Employee Safety Discipline Letter #2

Attn: Mary Anderson, Department 27

Dear Mary:

On January 5, 1998, you met with Steve Jones, your union representative; Bob Wilson, department supervisor; and me to discuss an unsafe act committed by you.

On January 3rd you were stacking boxes of widgets in department 27 while using a forklift. Another employee asked that you raise him up on the forks to reach some other boxes in the area and you did. Your department supervisor, Bob Wilson, observed this unsafe act.

It was agreed at the meeting that you had been given fork lift training, know the proper safety procedures and that you were in error in committing this unsafe act which could have resulted in an employee injury or fatality.

Your past conduct concerning safety in our company has been very good. Your attitude of taking responsibility and acknowledging this unsafe act has also been noted.

The purpose of this letter is to emphasize the importance of safe job practices. It is now your responsibility to prevent such unsafe acts from happening in the future. This letter is meant to correct the problem, and will only be referred to if further corrective action becomes necessary.

Yours truly,

Ronald Stevens
Personnel Director

Doctor's Modified Duty Cover Letter

Dear Doctor:

Our company provides transitional duty work for our injured employees.

The enclosed is a physical capability form consistently used by our company and its injured employees. Please fill out the form as specifically as you can, and send it back with our employee. Our employees cannot return to work without this slip from their doctor.

We are interested in knowing what an employee can do vs. what she/he cannot do.

Our policies/procedures/philosophy provide a work environment that supports an early return-to-work into transitional duty for all our injured employees. We believe that the sooner an employee is returned to work after an injury into a supportive and safe environment, the faster his/her recovery will be.

If you find that an employee is physically capable of doing anything at all, including less than a full day at work, we will have transitional duty work available for that employee.

Thank you for your positive involvement in our employee's recovery.

Sincerely,

Safety Director

Medical Capabilities Form

Employee:_____   Employer:_____

Employer Contact person:_____

Date of injury/illness:_____   Nature of injury/illness:_____

WORK STATUS

Diagnosis:_____ Work-related:_____Nonworkrelated:_____

Treatment:_____

Disposition:_____Return-to-work date (no limitations) :_____

Return-to-work date (with limitations) from:_____to:_____
Unable to work   from:_____to :_____

Return to clinic on:_____

Prognosis:_____

Referral: To Consultant - Doctor:_____   Date & Time:_____
Physical Therapy :_____   Frequency:_____

WORK RESTRICTIONS

Restrictions apply to: Work_____   Home_____   Leisure_____

During an 8-hour day, restrict:

____ Sitting   __1   __2   __3   __4   __5   __6   __7   ___8 (hours)

____Standing   __1   __2   __3   __4   __5   __6   __7   ___8 (hours)

____Walking   __1   __2   __3   __4   __5   __6   __7   ___8 (hours)

____Pushing   __1   __2   __3   __4   __5   __6   __7   ___8 (hours)

____Lifting/Carrying   ___ 0-10 lbs   ___ 11-24 lbs   ___ 25-34 lbs

___ 35-50 lbs   ___ 51-74 lbs   ___ 75-100 lbs

____Lifting frequency/number of times per hour

____Bending limit in degrees/frequency per hour

____Restricted use of hands   Right _____Left _____

____Gripping no./hr_____Grasping no./hr. _____Pinching no./hr._____Rotation no./hr. _____

____Restricted use of feet   Right_____   Left_____   Both_____

____Dry, clean work _____

____No operation moving equipment or machinery.

____No exposure to chemicals (specify)_____

____No static position (specify)_____

____No climbing/overhead work_____

____Other_____

Physician's comments:

Signature_____ Date_____

*Please call the employer contact person after the employee is examined.
*The employee is to return the completed form to the Safety Director.

Supervisor Safety Performance Evaluation

Name:_____   Job Title:_____

Department:_____

Date : _____   Date:_____

(this evaluation)   (last evaluation)

Evaluation of Performance on Present Position:

Poor	Below Average	Average	Above Average	Outstanding
1	**2**	**3**	**4**	**5**
Serious weakness to overcome	Meets some requirements	Performance meets overall requirements	Performance above average	Outstanding performance

Principal Accountabilities

Explanatory comments are essential where performance is noted to be provisional or marginal.

1. _____ Ensures compliance of subordinates' activities regarding safety rules and regulations.
2. _____ Actively enforces department safety procedures for department employees in a fair and consistent manner.
3. _____ Uses acquired safety knowledge to improve safety in the department.
4. _____ Maintains open and receptive communication with subordinates regarding department safety issues
5. _____ Is resourceful and uses initiative in resolving departmental safety problems.
6. _____ Is vigilant on a daily basis in observing, correcting, and/or reporting safety hazards.
7. _____ When required, completes employee/property incident investigations with thoroughness and accuracy in a timely manner.
8. _____ Has reduced recordable incident rate since last evaluation period.

A.   Plans for Employee Development:

_____
_____
_____
_____

B.   Attitude and commitment toward safety which, to a marked degree, either adds or detracts from overall performance:

_____
_____
_____

C.   Significant changes in performance, which have been noted since last review:

_____
_____
_____
_____

D.   Key areas where performance can be improved:

_____
_____
_____
_____

E.   Comments on differences of opinion concerning review:

_____
_____
_____
_____

This appraisal covers performance for the previous _____ months.
Appraisal form completed by_____   Date_____
Appraisal form received by_____   Date_____

I have reviewed the job description and it is

_____ current   _____ needs revision

Employee's Signature_____

Supervisor's Signature_____

Safety Committee Agenda

DATE:

TO: All Safety Committee Members

FROM: Safety Director

SUBJECT: Monthly Safety committee Agenda

On Tuesday, April 9, 1998 at 9:00 A.M. the Safety committee will meet in the conference room. Following the safety meeting, there will be a one-hour claims review meeting!

AGENDA

I. Roll Call

II. Introduction of Visitors

III. Team Safety Department of the Month.

IV. Old Business

    a. Top three departments with injury-free days

    b. *To be determined*

    c. Other old business

V. New Business

    a. *To be determined*

    b. Injury review

    c. Other new business

FOLLOWING THE SAFETY MEETING THERE WILL BE A ONE-HOUR CLAIMS REVIEW MEETING!

Safety Hazard/Suggestion Report

Report of Hazard:_____

Name:_____

Department:_____

Safety Suggestion:_____

Please state specifically what your hazard or suggestion is:

Date	
	Signature of Dept. Supervisor:

Safety Incentive Program

YOU ARE AN IMPORTANT MEMBER OF OUR SAFETY TEAM

Purpose of Program

The Employee Safety Incentive Award Program is designed to reward you, the employee, for safe work practices on an individual basis.

Eligibility

Because management believes that working safely is important to both the employee and the company, this program is open to all employees who worked at least 50 days in a calendar quarter without a recordable injury. A calendar quarter is defined as a three-month period ending March 31st, June 30th, September 30th, or December 31st. Work days include your normal 8-hour day, jury duty, union business, emergency leave, and vacation days.

Determining Your Participation

Your participation depends upon completing your work assignments during a quarter without a recordable injury. A recordable injury is any work injury that must be treated by a health care professional off the company property, whether it be our company medical consultant or your own physician.

Policy on Medical Treatment

The company's policy on medical treatment is as follows:

If you suffer a work injury that you feel should be examined by a health care professional, but it is not of an emergency nature, the company will make an appointment for you to see our company medical consultant. If your injury is of a serious emergency nature you will be transported immediately to the hospital. If you suffer a work injury the company may require that you see our medical consultant even though you feel that it is not necessary. This is for your protection as well as ours.

Employee Safety Incentive Awards

There are four award levels, all increasing in worth. Each level represents the value of the award you will receive:

Level 1	$35.00
Level 2	$75.00
Level 3	$125.00
Level 4	$200.00

Each quarter that you successfully complete without a recordable injury entitles you to an incentive award. You may receive your award at the end of the quarter or defer/bank this award until the next quarter. Completion of an entire year without a recordable injury will result in a Level 4 award. If you elect to bank your awards, and if you subsequently incur a recordable injury, you must forfeit all awards earned up to that point. You will then have to wait until the start of the next quarter before you become eligible to earn a Level 1 award.

SUGGESTED REFERENCES

Besides the hundreds of reference books available to you, these references to books will give you the basic "know how" and workable knowledge you will need in your role.

1. National Safety Council
    National Office Order Department
    PO Box 558
    Itasca, IL 60611-0933
    1-800-621-7619   24-hour FAX (708) 285-0797
    Call to ask for a copy of their General Materials Catalog. This is one of the best organizations you can use for all types of reference materials and training programs for yourself and your supervisors. The books you need to order first are:

    Accident Facts
    Supervisor Safety Manual
    Introduction to Occupational Health and Safety
    Remember that your local safety council is affiliated with the National Safety Council and can assist you on a local level.

2. Comprehensive Loss Management, Inc.
    7750 West 78th Street
    Minneapolis, MN. 55439
    612-944-1959

This firm is especially good at producing safety training material that keeps you in compliance with OSHA law. Training videos and booklets ranging from right-to-know to fork lift operations are clear, concise and produced for your workers to clearly understand. They also produce a fine supervisor training program. Their accident report forms are the best I have seen. I highly recommend this firm.

3. The Merritt Company's OSHA Reference Manual.

This two volume set is presented in an easy to understand format with self-inspection check list and monthly OSHA updates. Their telephone number is: 1-800-638-7597.

The Merritt Company
1661 Ninth Street
P.O. Box 955
Santa Monica, CA. 90406

4. Dupont Safety and Environmental Management
Training Materials
Montgomery Building, Room 208
Willmington, Delaware 19880-080

Dupont is one of the recognized world leaders in safety training material. I have used their S.T.O.P. (Safety Training Observation Program) and found it to be very successful in reducing losses. Dupont has many other outstanding training programs.

5. The Bureau of Business Practice:
Safety Compliance Letter with OSHA Highlights

Bureau of Business Practice
24 Rope Ferry Road
Waterford, CT. 06386
They also have other safety training material that you can use.

6. Safety Supply Vendors:
Your safety supply sales representative can be a source of great help to you. They are usually aware of required standards, and may be able to provide training.

GLOSSARY OF TERMS

abatement period. The time given to the employer to abate or correct the violation of standard.

accident investigation. The procedure used to determine the what, where, when, why, and how of an employee injury. The purpose of the investigation is to prevent the recurrence.

appeal. When a issued citation is issued by OSHA, the employer may request a meeting to discuss the case. Often this informal conference can be used to resolve differences and reach an equitable agreement on compliance issues.

citation. A statement from OSHA to the employer stating the violation of safety and health regulations.

claims report. Also called a loss run. A report from your insurance carrier showing how your company's dollars are being spent.

claims representative. The person from the insurance carrier who handles claims services for treatment, rehabilitation, and workers' compensation benefits for the injured employee.

de minimus. A violation that is a minor infraction of the standards. It is included on the OSHA inspection report, but no citations are issued.

EAP. Employee Assistance Program

EPA. Environmental Protection Agency

first line supervisor. The person in the company who has immediate supervision responsibilities for the work and safety of employees.

formal inspection. The Safety Director and members of management conduct a company-wide safety inspection.

imminent danger. An impending or threatening dangerous situation that could be expected to cause death or serious injury to persons in the immediate future unless corrective measures are taken.

incident rate. A measuring device for occupational injury and illness occurrences that companies can use to compare themselves against like companies.

informal inspection. A department supervisor and one or two employees conduct a safety inspection.

indemnity. Money the insurance company pays to employees for lost wages due to disabilities, rehabilitation or death.

insurance agent. A person who serves as the bridge between the company and the insurance carrier.

job instruction. Training provided by the supervisor for a new employee on how to perform work tasks safely.

KISS. Keep It Simple, Stupid.

loss control representative. The person from the insurance carrier who provide companies technical safety assistance.

loss run. Report from the insurance carrier telling you how your company dollars are being spent.

lost time injury. An occupational injury or illness that requires the employee to be away from work for treatment and/or rehabilitation.

lost work days. The number of days the employee is off work due to an occupational injury or illness.

modified duty. Work that an injured employee can do while they are on medical restrictions.

navigate coordinator. The company employee who is responsible for the navigate program and assisting employees with medical needs, workers' compensation insurance and return-to-work procedures.

navigate program. Company policies and procedures to assist injured employees in the recovery and return-to-work processes.

new employee orientation. Safety training given to each new employee before beginning work.

nine injury categories. The nine basic categories of injuries used on the OSHA log and insurance loss runs. The categories are: struck by, body mechanics, laceration/cut/tear/puncture, struck against, contact with temperature extremes, fall/slip/trip, caught-in-between, eye, and miscellaneous.

notice of contest. A document the employer files with OSHA to contest a citation.

NSC. National Safety Council

100% Rule. An employee remains off work until he or she is 100% recovered from an occupational injury or illnesses.

OSHRC. Occupational Safety and Health Review Commission

OSHA. Occupational Safety and Health Administration

OSHA compliance officer. A federal or state employee that inspects your company for compliance with OSHA standards.

OSHA inspection. A compliance officer from OSHA inspects a facility for violations of the OSHA standards.

OSHA standards. Rules and regulations that apply to safety.
  1910—General Industry
  1926—Construction
  1928—Agriculture
  1915, 1917, 1918—Maritime

OSHA 200 Log. A required form from the Occupational Safety and Health Administration to record occupational injuries and illness.

PMA. Petition for Modification of Abatement

PPE. Personal Protective Equipment

personal protective equipment. Items worn by the employee on the job for protection from hazards. Some examples of PPE are: safety glasses, respirators, gloves, hard hats, and safety shoes.

premium. The amount of money the insurance company charges a company for insurance coverage.

professional medical consultant. Physicians that treat occupational injuries and illnesses for the company.

QRC. Qualified Rehabilitation Consultant

RAACK. Acronym for responsibility, authority, accountability and knowledge.

recordable injury. Any injury or illness that goes on the OSHA log.

repeat violation. If the company has been inspected and a violation discovered, and upon reinspection, the same violation or a similar one is found.

return-to-work. Policies and procedures used by a company to return an employee to work while the employee is recovering from an occupational injury or illness.

risk. The possibility of harm or loss.

safety. Freedom from danger, risk, or injury.

Safety Director. The designated person in the company to develop, implement and maintain an occupational safety and health program.

safety inspection. A walk-through of the facility to determine hazardous conditions that could cause injury or death to an employee or customer.

safety policy. A statement of the company's position on safety and procedures to manage safety within the company.

serious violation. A violation that has the potential of causing a serious injury or death, and it is obvious that the company should have known of the hazard.

SIC. Standard Industrial Classification code.

wwllful violation. When OSHA determines that you knew there was a hazardous condition that could kill or cause serious injury and you didn't correct it.

workers' compensation. Insurance coverage for employees in the event they are injured and/or become ill in and out of the course of employment.

BIBLIOGRAPHY

Bureau of Business Practice. *BBP Standard Manual for Supervisors.* Waterford, CT: Prentice Hall, 1990.

Benson, Tracy E. "Intangibles, The Real Bottom Line," *Industry Week*, August 16, 1993: 19.

Benson, Tracy E. "Paul O'Neill, True Innovation, True Values, True Leadership," *Industry Week,* April 19, 1993: 25.

"Control Comp Costs with a Team Approach," *Safety Compliance Letter—with OSHA Highlights*, Bureau of Business Practice, no. 2119, October 10, 1993: 1.

Dennis Leslie E. and Onion, Meredith L. *Out In Front: Effective Supervision in the Workplace.* Itasca, IL: National Safety Council, 1990.

Ferry, Ted S. *Modern Accident Investigation and Analysis, Second Edition.* New York: John Wiley & Sons, 1988.

"It's a Pain in the Back," *The Pryor Report*, November 1993: 8.

Mansdorf, S.Z. *Complete Manual of Industrial Safety.* Englewood Cliffs, NJ: Prentice Hall, 1993.

National Safety Council. *Accident Facts, 1997 Edition.* Chicago, IL, National Safety Council, 1994.

— *Accident Prevention Manual for Business & Industry, 10th Edition.* Itasca, IL: National Safety Council, 1996.

— *Product Safety: Management Guidelines.* Itasca, IL: National Safety Council, 1989.

— *Protecting Workers Lives, 2nd Edition: A Safety and Health Guide for Unions.* Itasca, IL: National Safety Council, 1992.

— *Supervisors' Safety Manual, 8th Edition.* Itasca, IL: National Safety Council, 1993.

"NSC Publishes Safety Committee Survey," *OSHA News,* November 29, 1993: 1.

"Physical and Occupational Therapists Essential for Safety Program: An Interview with Michael V. Manning, ASP," *Work Injury Management News and Digest,* March/April, 1993: 10.

Pierce, F. David. *Total Quality for Safety and Health Professionals.* Rockville, MD: Government Institutes, Inc., 1995.

The Merritt Company. *OSHA Reference Manual, Occupational Safety and Health Compliance Simplified.* Santa Monica, CA: The Merritt Company.

Thompson, Roger. "Taking Charge of Workers" Comp", *Nation's Business,* October, 1993: 18.

Verespej, Michael A. "Better Safety Through Empowerment," *Industry Week,* November 15, 1993: 56.

— "Roger Meade: Running on People Power," *Industry Week,* October 18, 1993: 13.

Walsh, James. *"Taking Control Workers' Comp for Employers: How to Cut Claims, Reduce Premiums, and Stay Out of Trouble."* Santa Monica, CA: Merritt Publishing, 1993.

"When 'Going Bare' Means Gaining Control," *Merritt Workers' Comp News,* November 22, 1993: 4.

Will, Robert J. *How to Legally Beat Workers' Compensation, Before It Beats You.* Minneapolis, MN: Rate Consultants, Inc., 1989.

Index

Government Institutes Mini-Catalog

PC #	ENVIRONMENTAL TITLES	Pub Date	Price
629	ABCs of Environmental Regulation: Understanding the Fed Regs	1998	$49
627	ABCs of Environmental Science	1998	$39
585	Book of Lists for Regulated Hazardous Substances, 8th Edition	1997	$79
579	Brownfields Redevelopment	1998	$79
4088	CFR Chemical Lists on CD ROM, 1997 Edition	1997	$125
4089	Chemical Data for Workplace Sampling & Analysis, Single User Disk	1997	$125
512	Clean Water Handbook, 2nd Edition	1996	$89
581	EH&S Auditing Made Easy	1997	$79
587	E H & S CFR Training Requirements, 3rd Edition	1997	$89
4082	EMMI-Envl Monitoring Methods Index for Windows-Network	1997	$537
4082	EMMI-Envl Monitoring Methods Index for Windows-Single User	1997	$179
525	Environmental Audits, 7th Edition	1996	$79
548	Environmental Engineering and Science: An Introduction	1997	$79
643	Environmental Guide to the Internet, 4th Edition	1998	$59
560	Environmental Law Handbook, 14th Edition	1997	$79
353	Environmental Regulatory Glossary, 6th Edition	1993	$79
625	Environmental Statutes, 1998 Edition	1998	$69
4098	Environmental Statutes Book/CD-ROM, 1998 Edition	1997	$208
4994	Environmental Statutes on Disk for Windows-Network	1997	$405
4994	Environmental Statutes on Disk for Windows-Single User	1997	$139
570	Environmentalism at the Crossroads	1995	$39
536	ESAs Made Easy	1996	$59
515	Industrial Environmental Management: A Practical Approach	1996	$79
510	ISO 14000: Understanding Environmental Standards	1996	$69
551	ISO 14001: An Executive Report	1996	$55
588	International Environmental Auditing	1998	$149
518	Lead Regulation Handbook	1996	$79
478	Principles of EH&S Management	1995	$69
554	Property Rights: Understanding Government Takings	1997	$79
582	Recycling & Waste Mgmt Guide to the Internet	1997	$49
603	Superfund Manual, 6th Edition	1997	$115
566	TSCA Handbook, 3rd Edition	1997	$95
534	Wetland Mitigation: Mitigation Banking and Other Strategies	1997	$75

PC #	SAFETY and HEALTH TITLES	Pub Date	Price
547	Construction Safety Handbook	1996	$79
553	Cumulative Trauma Disorders	1997	$59
559	Forklift Safety	1997	$65
539	Fundamentals of Occupational Safety & Health	1996	$49
612	HAZWOPER Incident Command	1998	$59
535	Making Sense of OSHA Compliance	1997	$59
589	Managing Fatigue in Transportation, *ATA Conference*	1997	$75
558	PPE Made Easy	1998	$79
598	Project Mgmt for E H & S Professionals	1997	$59
552	Safety & Health in Agriculture, Forestry and Fisheries	1997	$125
613	Safety & Health on the Internet, 2nd Edition	1998	$49
597	Safety Is A People Business	1997	$49
463	Safety Made Easy	1995	$49
590	Your Company Safety and Health Manual	1997	$79

Government Institutes

4 Research Place, Suite 200 • Rockville, MD 20850-3226
Tel. (301) 921-2323 • FAX (301) 921-0264
Email: giinfo@govinst.com • Internet: http://www.govinst.com

Please call our customer service department at (301) 921-2323 for a free publications catalog.

CFRs now available online. Call (301) 921-2355 for info.

GOVERNMENT INSTITUTES ORDER FORM

4 Research Place, Suite 200 • Rockville, MD 20850-3226
Tel (301) 921-2323 • Fax (301) 921-0264
Internet: http://www.govinst.com • E-mail: giinfo@govinst.com

3 EASY WAYS TO ORDER

1. Phone: **(301) 921-2323**
Have your credit card ready when you call.

2. Fax: **(301) 921-0264**
Fax this completed order form with your company purchase order or credit card information.

3. Mail: **Government Institutes**
4 Research Place, Suite 200
Rockville, MD 20850-3226 USA
Mail this completed order form with a check, company purchase order, or credit card information.

PAYMENT OPTIONS

❑ **Check** (*payable to Government Institutes in US dollars*)

❑ **Purchase Order** (*This order form must be attached to your company P.O. Note: All International orders must be prepaid.*)

❑ **Credit Card** ❑ VISA ❑ MasterCard ❑ AMERICAN EXPRESS

Exp.___/____

Credit Card No. _____

Signature _____

(Government Institutes' Federal I.D.# is 13-2695912)

CUSTOMER INFORMATION

Ship To: (Please attach your purchase order)

Name: _____

GI Account # (*7 digits on mailing l...*)

Company/Institution: _____

Address: _____
(Please supply ...

City: _____ St...

Zip/Postal Code: _____

Tel: ( ) _____

Fax: ( ) _____

Email Address: _____

Bill To: (if different from ship-to address)

...e supply street address for UPS shipping)

State/Province: _____

Country: _____

Qty.	Product Code		Price

Subtotal _____
Residents add 5% Sales Tax _____
...nd Handling (see box below) _____
Total Payment Enclosed _____

❑ **New Edition No Obliga...**
Please enroll me in this program fo...
Institutes will notify me of new editi...
that there is no obligation to purch...
reminder that a new edition has be...

15 DAY MONEY...
If you're not completely satisfied wi...
15 days for a full and immediate re...

Outside U.S:
Add $15 for each item (Airmail)
Add $10 for each item (Surface)

Now Order Onli... ...RCE CODE: **BP01**